Biological Substrates of Human Sexuality

Biological Substrates of Human Sexuality

edited by Janet Shibley Hyde

AMERICAN PSYCHOLOGICAL ASSOCIATION
WASHINGTON, DC

Published by
American Psychological Association
750 First Street, NE
Washington, DC 20002
www.apa.org

To order
APA Order Department
P.O. Box 92984
Washington, DC 20090-2984
Tel: (800) 374-2721
Direct: (202) 336-5510
Fax: (202) 336-5502
TDD/TTY: (202) 336-6123
Online: www.apa.org/books/
E-mail: order@apa.org

In the U.K., Europe, Africa, and the Middle East, copies may be ordered from
American Psychological Association
3 Henrietta Street
Covent Garden, London
WC2E 8LU England

Typeset in Goudy by World Composition Services, Inc., Sterling, VA

Printer: Victor Graphics, Inc., Baltimore, MD
Cover Designer: Berg Design, Albany, NY
Technical/Production Editor: Gail B. Munroe

The opinions and statements published are the responsibility of the authors, and such opinions and statements do not necessarily represent the policies of the American Psychological Association.

Library of Congress Cataloging-in-Publication Data

Biological substrates of human sexuality / edited by Janet Shibley Hyde.— 1st ed.
    p. cm.
  Includes bibliographical references and index.
  ISBN 1-59147-250-4
  1. Sex (Biology) I. Hyde, Janet Shibley.

QP251.B5626 2005
612.6—dc22

2004026914

British Library Cataloguing-in-Publication Data
A CIP record is available from the British Library.

Printed in the United States of America
First Edition

CONTENTS

Contributors .... *vii*

Acknowledgments .... *ix*

Chapter 1. Introduction: The Role of Modern Neuroscience in
Sexuality Research 3
Janet Shibley Hyde

Chapter 2. The Genetics of Sexual Orientation 9
Janet Shibley Hyde

Chapter 3. The Role of Hypothalamus and Endocrine
System in Sexuality 21
Dick F. Swaab

Chapter 4. The Central Control and Pharmacological
Modulation of Sexual Function 75
Kevin E. McKenna

Chapter 5. Brain Activity Imaging During Sexual Response in
Women With Spinal Cord Injury 109
Barry R. Komisaruk and Beverly Whipple

Chapter 6. Neuroanatomical and Neurotransmitter Dysfunction
and Compulsive Sexual Behavior 147
Eli Coleman

Chapter 7. Synthesis: Overarching Themes and Future
Directions for Research 171
Janet Shibley Hyde and Beverly Whipple

Author Index 179

Subject Index 195

About the Editor 207

CONTRIBUTORS

Eli Coleman, University of Minnesota Medical School, Minneapolis
Barry R. Komisaruk, Rutgers University, Medford, NJ
Janet Shibley Hyde, University of Wisconsin—Madison
Kevin E. McKenna, Northwestern University, Chicago, IL
Dick F. Swaab, Netherlands Institute for Brain Research, Amsterdam, the Netherlands
Beverly Whipple, Rutgers University, Medford, NJ

ACKNOWLEDGMENTS

This volume began with a conference on the biological substrates of human sexuality at the beginning of a meeting of the Society for the Scientific Study of Sexuality in June 2002. We are grateful to those who provided funding for the conference and the writing of these chapters. The Science Directorate of the American Psychological Association supported the conference in the form of funds for graduate students to attend the conference, providing them with exciting opportunities to interact with the premier scientists in the field. Funding for speakers was provided by the National Institute of Child Health and Human Development and by the Sexuality Research Program of the Social Science Research Council (Diane DiMauro, program officer).

Biological Substrates of Human Sexuality

1

INTRODUCTION: THE ROLE OF MODERN NEUROSCIENCE IN SEXUALITY RESEARCH

JANET SHIBLEY HYDE

The past decade has witnessed an explosion of neuroscience research on sexuality. This research has been driven by three forces: the quest of drug companies for Viagra and even better successors; technological advances in numerous areas, including functional magnetic resonance imaging (fMRI) technology; and advances in neuroscience techniques in both animal and human postmortem research. The result is that a tremendous amount more is known today about the biological substrates of human sexuality than 10 years ago.

I use the terms *neuroscience* and *biological substrates* in their broadest sense here, to include central and peripheral neural pathways, neurotransmitters, the endocrine system, genetic blueprints, and interactions among all these systems. The emphasis on biological substrates is not intended to slight the enormous cultural, interpersonal, and individual life history influences on sexuality, which are well documented elsewhere (e.g., Adkins-Regan, 2002; Frayser, 1985, 1994; Gregersen, 1996; Herdt, 1984, 1997; Wyatt, 1997). I seek only to focus on a specific piece of the puzzle, namely the biological substrates.

This volume is above all integrative. It integrates research on humans and animals; research on males and females; research using vastly different methodologies including twin studies, fMRI, viral tracers to identify neuroanatomy, and experimental pharmacology; and research at multiple levels of analysis from the molecular (McKenna's chapter on pharmacology) to the familial (Hyde's chapter on family resemblance for sexual orientation). It integrates a fascinating array of aspects of sexuality, including sexual orientation, orgasm, erectile dysfunction, the sexual functioning of women with spinal cord injury, and compulsive sexual behavior. The authors represent a variety of disciplines, including psychology, physiology, neuroscience, urology, and nursing.

This book will be useful to many sectors of the community of psychologists. Practitioners increasingly need to understand the biological substrates of both male and female sexual functioning and dysfunction; for example, they need to know what sorts of clients might benefit from Viagra. Those who teach human sexuality, behavioral neuroscience, or biopsychology in colleges and universities will find in this volume much-needed information for updating their teaching. And certainly researchers in sexuality and related fields will appreciate the cutting-edge reviews found in the chapters that follow.

This book integrates multiple levels of analysis, ranging from the most basic to the most complex. The book begins with genes, the basic biological blueprints, and Hyde's chapter on the genetics of sexual orientation. The next chapter moves to the cellular level, focusing on the hypothalamus, a region of the brain that has been implicated particularly in sexual behavior and identity. Swaab's chapter reviews research on the important regions of the human hypothalamus and their connections to the sex hormone system, also discussing data from animal research. McKenna focuses on the central nervous system, synthesizing research that has identified the pathways in the brain and spinal cord, and the neurotransmitters that are involved in sexual function and dysfunction. Komisaruk and Whipple review their pathbreaking research using fMRI and positron emissions tomography (PET) scans to understand how women with complete spinal cord injury can nonetheless feel genital stimulation and experience orgasm. Finally, we move to the level of clinical applications, with Coleman's chapter on compulsive sexual behavior and the neural dysregulation that may underlie it.

One overarching theme is prominent across chapters. Sexuality is multifaceted and complex, even in the rat. In the male rat, for example, mounting, intromission with erection, and ejaculation are all distinct processes with somewhat different neural control. In humans, sexual orientation may be defined by behavior, attraction, or identity, and individuals may be discordant from one aspect to another. It is unfortunate that some recent trends in the field ignore these distinctions. For example, some speak of "female sexual dysfunction" as if it were a single diagnostic category, masking the multiplicity of sexual dysfunctions and dissatisfactions that women can experience. Each of those dysfunctions may have distinct biological substrates and sources of learning through the environment and experience. The chapters that follow give ample recognition to the complexity of sexuality.

In chapter 2, Hyde reviews the evidence on whether there is a *genetic basis for sexual orientation*. Much of the evidence comes from studies of identical (monozygotic or MZ) twins compared with fraternal (dizygotic or DZ) twins and their concordance for homosexuality—or, more precisely, nonheterosexual orientation, including both homosexuality and bisexuality. Sufficient studies using strong methods now exist to support the conclusion that nonheterosexual orientation is moderately heritable. Other researchers have advanced the hypothesis that homosexuality in some men is influenced by a gene on the X chromosome, called Xq28. The replicability of these findings is somewhat ambiguous. Results of the first genomewide scan for male sexual orientation, published in 2005, are included. Hyde also reviews genomic research with other species such as fruit flies, which has identified specific genes controlling specific aspects of sexual behavior such as the courtship ritual. Finally, Hyde considers the ethical issues that will arise if specific genes influencing homosexual orientation are identified.

Swaab has conducted a long and distinguished program of research on the *hypothalamus and endocrine system* and reports on this and related research in chapter 3. Patterns of sex differences and control of sexual behavior have been identified in certain regions of the hypothalamus. For example, the sexually dimorphic nucleus of the preoptic area (SDN-POA, also known as the interstitial nucleus of the anterior hypothalamus, or INAH-1 in humans) shows marked anatomical differences between males and females across numerous species. Lesions of the preoptic area eliminate mounting, intromission, and ejaculation in male rats. This region also contains an abundance of both androgen receptors and estrogen receptors and contributes to the regulation of circulating levels of these sex hormones throughout the body. Swaab's discussion ranges over questions of sexual orientation and transsexualism and the possible involvement of the hypothalamus in these patterns of behavior and identity. A glossary beginning on p. 73 makes this technical chapter more accessible. Clearly the hypothalamus is one of the brain's real hot spots for sexual functioning.

In chapter 4, McKenna considers the role of the *central nervous system—* brain and spinal cord—and neurotransmitters, focusing particularly on mechanisms influencing erection in males. McKenna has been a major contributor to the basic science that supported the development of prescription drugs to treat erectile dysfunction. Viagra acts peripherally, facilitating vasocongestion in the penis. Strategies for successor drugs are capitalizing on central mechanisms as well. McKenna reviews new discoveries about the neuroanatomy of sexual response, focusing on sites such as reflex centers in the spinal cord, and the inhibitory role of the brainstem (medulla and pons). In his consideration of the pharmacology of sexual response, McKenna focuses particularly on the neurotransmitter serotonin, which generally has an inhibitory effect on sexual functioning. This leads to the unfortunate result that many drugs prescribed for the treatment of depression and designed to raise serotonin levels—particularly the selective serotonin reuptake inhibitors (SSRIs)—have marked sexual dysfunctions as side effects. McKenna considers several other neurotransmitters as well, including dopamine, noradrenaline, oxytocin, and nitric oxide.

Given that brain and spinal cord mechanisms of sexual response (as reviewed by McKenna) are well established, Komisaruk and Whipple begin chapter 5 with a paradox: How could *women with complete spinal cord injury* feel genital stimulation and experience orgasm? For genital sensations to be consciously perceived, they must travel from sensory neurons in the genitals, to neurons in the spinal cord, and then to an ascending pathway through the spinal cord that eventually registers in the cortex. How could women whose spinal cords have been completely severed at a level above the point where the sensory neurons join the spinal cord feel genital sensation when the ascending pathway has been severed? Komisaruk and Whipple's pathbreaking research began by taking the reports of these women seriously rather than dismissing them as "phantom" sensations. Through a remarkable series of laboratory studies of these women, using fMRI and PET technologies, they have demonstrated an alternative sensory pathway involving the Vagus nerve, which completely bypasses the spinal cord and projects directly to the brain. Their imaging studies have also identified a region of the hypothalamus that is activated during orgasm. Their findings have profound implications for practitioners who may mistakenly tell their clients with spinal cord injury that sex is a thing of the past for them. Komisaruk and Whipple show how and why sexual response may still be possible in many cases of spinal cord injury.

In chapter 6, Coleman addresses the issue of *compulsive sexual behavior* (CSB, also called sexual addiction by some) and its substrates in neuroanatomy, hormones, and neurotransmitters. Coleman is arguably the nation's expert on CSB and its treatment. He finds that the limbic system of the brain is particularly implicated. Although findings regarding elevated levels of testosterone in individuals with CSB are equivocal, there is substantial evidence that antiandrogen treatments are effective. Posttraumatic stress disorder and the associated neural consequences may be implicated in some cases of people with CSB who were victims of childhood sexual abuse. SSRIs and other antidepressants are effective in treating many cases of CSB.

In the final chapter, Hyde and Whipple synthesize the findings reported in the previous chapters and identify five overarching themes: Biological substrates are developmental; sexuality is complex and multifaceted; certain regions of the brain, particularly the hypothalamus, are clearly linked to sexuality; the spinal cord plays a key role as well; and the biological substrates consist of multiple, interlocking systems, including genes, the brain, and the endocrine system. They conclude with suggestions about future directions for research on the biological substrates of human sexuality.

The array of research displayed in these chapters is dazzling. It testifies to the possibility of rapid scientific progress in understanding crucial areas of human behavior such as sexuality. This research has flourished despite societal taboos regarding sex research and opposition of some in the federal government. We hope that readers will gain much from reading these chapters.

REFERENCES

Adkins-Regan, E. (2002). Development of sexual partner preference in the zebra finch: A socially monogamous, pair-bonding animal. *Archives of Sexual Behavior*, *31*, 27–33.

Frayser, S. G. (1985). *Varieties of sexual experience: An anthropological perspective on human sexuality*. New Haven, CT: Human Relations Area Files Press.

Frayser, S. G. (1994). Defining normal childhood sexuality: An anthropological approach. *Annual Review of Sex Research*, *5*, 173–217.

Gregersen, E. (1996). *The world of human sexuality: Behaviors, customs, and beliefs*. New York: Irvington.

Herdt, G. H. (1984). *Ritualized homosexuality in Melanesia*. Berkeley: University of California Press.

Herdt, G. H. (1997). *Same sex, different cultures*. Boulder, CO: Westview Press.

Wyatt, G. E. (1997). *Stolen women: Reclaiming our sexuality, taking back our lives*. New York: Wiley.

2

THE GENETICS OF
SEXUAL ORIENTATION

JANET SHIBLEY HYDE

We live in the era of the Human Genome Project, and genes are being identified that control everything from full fruit development in strawberries to breast cancer susceptibility in women (Marshall, 1999; Miki et al., 1994). At the same time, this is an era in which issues of sexual orientation have been placed front and center. In 2003 alone, an openly gay man, the Rev. Gene Robinson, was confirmed as a bishop in the Episcopal Church, and an amendment was proposed to the U.S. Constitution specifying that marriage could occur only between a man and a woman. It is therefore not surprising that the lay public, scientists, and practitioners should wonder about the causes of sexual orientation and, specifically, whether there is a genetic basis for sexual orientation.

In this chapter I review the evidence on genetic influence on sexual orientation in humans, including evidence from twin and adoption studies, as well as evidence from studies using animal models. (For earlier reviews, see Bailey & Pillard, 1995; Mustanski, Chivers, & Bailey, 2002; Pillard & Bailey, 1998.) I conclude with a consideration of the ethical issues that are raised by this newly emerging science.

As we consider the evidence, an important methodological issue is the measurement of sexual orientation. Alfred Kinsey championed a behavioral definition captured in his famous 7-point rating scale in which 0 represented a person *all of whose sexual contacts were heterosexual*, 3 represented *equal amounts of heterosexual and homosexual experience*, and 6 represented *all same-gender sexual experience* (Kinsey, Pomeroy, & Martin, 1948; Kinsey, Pomeroy, Martin, & Gebhard, 1953). Identity, fantasy, and attraction are additional components of sexual orientation that contemporary researchers consider. *Sexual identity* refers to one's own self-label as being heterosexual, homosexual, bisexual, or other categories, such as queer. Identity typically is measured categorically. Moreover, sexual identity can be notably discordant from sexual behavior (Lever, Kanouse, & Rogers, 1992). For example, a woman may self-label as bisexual even though all her sexual experiences have been with men. Likewise, a woman may have experienced sexual attraction to other women but never acted on it. The well-known National Health and Social Life Survey, for example, asked respondents about behavior, identity, and attraction (Laumann, Gagnon, Michael, & Michaels, 1994). The measurement of sexual orientation is not standardized in genetic research, so differences in measurement may account for variations in outcomes across studies.

EVIDENCE FROM TWIN AND ADOPTION STUDIES

Behavior geneticists who study humans cannot perform the experimental manipulations that are possible when studying nonhumans. Therefore, they capitalize on a number of "natural" experiments, such as identical twins. Monozygotic (MZ), or identical, twins are two individuals with identical genotypes. The question, then, is whether they are concordant or discordant for the trait under study. In this paradigm, probands are identified who possess the trait under investigation—in this case, who are gay or lesbian—and are also a member of an MZ twin pair. The other member of the pair is then located and the sexual orientation of that individual is ascertained. If the other member of the pair also is gay, then the pair is said to be *concordant.* If not, then the pair is discordant. Generally results from these studies are reported as concordance rates—in this case, the percentage of the twin pairs in which both members are gay.

The study of identical twins by itself is limited because the twins not only share all their genes but also grew up in the same environment, so concordance might be due either to genetic factors or to environmental ones. To address this limitation, human behavior geneticists have used a variety of methods, including studies of MZ twins reared apart. For a relatively uncommon trait such as homosexuality, studies of identical twins reared apart are generally not feasible; one would have to locate people who were gay, who were members of an identical twin pair, and who were separated and reared apart from the first year of life. The resulting sample size would be tiny and would not permit statistical analysis.

The alternative, more feasible design is to compare concordance rates for identical twins with those from dizygotic (DZ) or nonidentical, fraternal twins. DZ twins are two individuals born at the same time who are not genetically identical but rather are as similar as siblings. Like identical twins reared together, DZ twins share a common environment. Therefore, any difference in concordance rates between MZ and DZ twins should be due to the greater genetic similarity of MZ twins.

Yet another design involves the study of adoptive siblings. They are genetically unrelated, but they are reared in the same environment. High concordance rates would testify to the importance of environment.

Early genetic studies of homosexuality by Kallmann (1952a, 1952b) indicated a strong genetic influence. Kallmann found perfect concordance for homosexuality among all the MZ pairs he studied, all of whom were men, and nonsignificant concordance among the DZ pairs. The methods of the study were later criticized, though, and one study failed to replicate it (Heston & Shields, 1968). For a time, the hypothesis of genetic influence on sexual orientation was discarded.

The modern era of behavior–genetic research on sexual orientation was ushered in by a series of studies by Bailey and colleagues. In the first study, they identified 56 gay men who were members of an MZ twin pair (Bailey & Pillard, 1991). When the identical twin brothers were located, the concordance rate proved to be 52%. In the comparison sample of 54 gay men who were members of a DZ twin pair, the concordance rate was only 22%. That the MZ twins were considerably more likely to be concordant for homosexuality is taken as evidence of genetic influence on sexual orientation. Nonetheless, the concordance rate for the MZ pairs was not 100%, indicating that genes do not completely control sexual orientation. Measurement of sexual orientation of the probands was self-definition; recruiting advertisements asked for volunteers who were gay or bisexual who had a twin. Sexual orientation was then confirmed during an interview using the Kinsey rating scale for both fantasy and behavior.

In this same study, Bailey and Pillard (1991) also located gay men who were adopted or had an adopted sibling. The concordance rate for these genetically unrelated pairs was 11%—far below the rate for MZ or DZ twins, and not much higher than one would expect at random, given that roughly 6% to 8% of men in the United States are gay (Hyde & DeLamater, 2003; Laumann et al., 1994).

In a companion study, Bailey, Pillard, Neale, and Agyei (1993) repeated the design with lesbians as probands. The concordance rates were 48% for MZ twin pairs (based on 71 pairs), 16% for DZ twin pairs (based on 37 pairs), and 6% for adoptive sisters (based on 35 pairs). That is, the results were very similar to those for gay men and indicated some genetic influence on women's sexual orientation, but not complete genetic control of sexual orientation.

Two other studies using the same design comparing MZ and DZ pairs have also appeared, although they have smaller sample sizes and combined male and female pairs. Whitam, Diamond, and Martin (1993) found a 66% concordance rate for MZ pairs and 30% for DZ pairs. They assessed sexual orientation using respondents' self-ratings of behavior on the 7-point Kinsey scale. King and McDonald (1992) found a concordance rate of 25% for MZ and 12% for DZ pairs, pooling male pairs and female pairs. Measurement was weak because cotwins were not contacted; the researchers simply accepted the reports of probands about their own and their twin's orientation.

The design used in these studies, although illuminating, contains some embedded biases. In particular, when probands are recruited, concordant pairs are considerably more likely to be identified than discordant pairs. If recruiting signs are placed in gay bars and newsletters for gays, either member of concordant pairs may see the ad and respond to it, whereas in discordant pairs, only one member of the pair can respond.

A design that corrects for this bias is the registry study. Australia, for example, has a twin registry, a national database containing roughly 25,000 twin pairs in Australia (Bailey, Dunne, & Martin, 2000). Registration is voluntary, which may introduce some bias, but for the purposes of sexual orientation research, the important point is that it is a general twin registry. That is, people were not recruited on the basis of sexual orientation or any other specific characteristic. Bailey and colleagues capitalized on this registry to identify nonheterosexual twins, that is, twins who scored ≥ 2 on the Kinsey scale. For men, the concordance rate for MZ twins was 20%, compared with 0% for DZ twins. For women, concordance was 24% for MZ twin pairs and 10.5% for DZ twins. The results are again consistent with a genetic model of sexual orientation. As expected on the basis of methodological considerations, the concordance rate was lower than that found in the earlier studies that used different recruiting methods. The superiority of the twin registry method means that the 20% to 24% concordance rate is likely to be more accurate.

A second registry study used the Minnesota Twin Registry and contacted never-married twins on the basis of the idea that they were most likely to be gay (Hershberger, 1997). That assumption, of course, ignores people who are bisexual and heterosexually married. Sexual orientation was assessed with distinct variables focusing on attraction, behavior, and identity. Concordance rates were higher for MZ lesbians (0.55 for attraction, 0.50 for behavior, and 0.52 for identity, based on 101 pairs) than for MZ gay men (0.23, 0.14, and 0.16, respectively, based on 76 pairs).

A U.S. national sample of twin pairs, again not recruited on the basis of sexual orientation, also provides evidence of genetic influence. The sample was drawn from the MacArthur Foundation's Midlife Development in the United States (MIDUS) survey. Kendler, Thornton, Gilman, and Kessler (2000) found a concordance rate for nonheterosexual orientation of 31.6% for MZ pairs (males and females combined, 19 pairs), compared with a 13.3% concordance for DZ twins (24 pairs). These results are consistent with those of Bailey et al. (2000) with the Australian twin registry. The MIDUS survey measured sexual orientation with a single item: "How would you describe your sexual orientation? Would you say you are heterosexual (sexually attracted only to the opposite sex), homosexual (sexually attracted only to your own sex), or bisexual (sexually attracted to both men and women)?" This item assesses a combination of identity and attraction, not behavior. In statistical analyses, the second and third response alternatives were combined so that two categories remained: heterosexual and nonheterosexual.

TWIN STUDIES: THE EQUAL ENVIRONMENTS ASSUMPTION

In twin studies, differences between concordance rates for MZ versus DZ twins are attributed to the greater genetic similarity of the MZ twin pairs. This logic relies on the assumption that DZ twins' environments are similar to each other to the same degree that MZ twins' environments are similar (Bailey & Pillard, 1995). The objection that has been raised is that parents might treat MZ twins more similarly than they do DZ twins. (Generally the sample of DZ twins is limited to same-sex pairs so that gender differences in behavior or in parents' treatment do not cloud the results; all MZ pairs, of course, are same-sex.) Empirical tests of the assumption, in the context of sexual orientation research, have generally supported the equal environments assumption. Bailey et al. (2000) found that concordant MZ twins were no more similar than discordant MZ twins on an index of childhood experiences. Kendler et al. (2000) found that twin resemblance for sexual orientation was not significantly predicted by measures of similarity of childhood environments.

OTHER FAMILIAL STUDIES

Other behavior–genetic designs involve assessing resemblance for other family relationships such as siblings. In one study, 6.4% of the sisters of lesbians were themselves nonheterosexual, compared with 0%

nonheterosexuals among sisters of heterosexual women (Bailey & Bell, 1993). In that same study, 9% of the brothers of homosexual men were themselves nonheterosexual. In another study, 15.4% of sisters of lesbian probands rated themselves homosexual or bisexual, as did 6.1% of brothers of male homosexual probands (Bailey & Benishay, 1993). These rates were higher than the rates of homosexuality reported by siblings of heterosexual probands: 3.5% for women and 0% for men. The results of these studies are also consistent with the hypothesis that sexual orientation is heritable, although the twin designs provide somewhat more persuasive evidence.

THE SINGLE-GENE HYPOTHESIS

Most behaviors that are genetically influenced, particularly those that are continuous traits with normal distributions, are thought to be polygenic, that is, influenced by multiple genes at multiple loci. Although the twin studies reviewed above are silent on this issue, it seems likely that the high but uneven concordance rates in MZ twins reflect polygenic influence.

In contrast to polygenic models are single-gene models, with the assumption that a single gene at a single locus can influence complex behavior. An example of this model is the much-publicized work of Hamer, Hu, Magnuson, Hu, and Pattatucci (1993). They used pedigree and linkage analyses to study the families of 114 homosexual men. *Pedigree analysis* refers to tracing the presence of a trait among relatives. *Linkage analysis* is done with samples of DNA from relatives and refers to determining whether the same alleles are present in a certain chromosome region (in this case, a region of the X chromosome) for relatives who are concordant for the trait. Men receive their X chromosome from their mother, so if a gene on the X chromosome produces a homosexual orientation, then rates of homosexuality should be elevated among men on the gay proband's mother's side of the family but not the father's side. The results indicated increased rates of homosexual orientation in maternal uncles of the probands but not in paternal uncles, suggesting transmission by a gene on the X chromosome. Linkage studies indicated inheritance through a gene on a particular region of the X chromosome, labeled Xq28.

A later study confirmed these results for Xq28 for homosexual orientation in males but not in females (Hu et al., 1995). Another study by an independent group, however, failed to replicate the findings for Xq28 (Rice, Anderson, Risch, & Ebers, 1999). The reasons for the inconsistency of findings is not entirely clear but may have to do with the protocol used for excluding cases. For example, Hamer et al. (1993) excluded families if the father was gay, whereas Rice and colleagues did not. Conclusions must remain tentative until there is further evidence of replication.

A major breakthrough came in 2005 with the first full genome scan for sexual orientation in men (Mustanski, DuPree, Nievergelt, Bocklandt, Schork, & Hamer, 2005). The sample included 456 individuals from 146 distinct families, all of which had at least two gay brothers. Mustanski and colleagues found evidence for genetic transmission by regions on chromosomes 7, 8, and 10. The strongest evidence was for chromosome 7, with equal contributions from maternal and paternal alleles. This region of chromosome 7 contains the gene for vasoactive intestinal peptide (VIP) receptor type 2 (VIPR2). In studies with mice, VIPR2 has been demonstrated to be essential for the development of the suprachiasmatic nucleus (SCN) of the hypothalamus. The SCN has previously been associated with sexual orientation in humans (Swaab, chap. 3 this volume). Therefore, the evidence of genetic linkage on that region of chromosome 7 is quite intriguing.

HOMOSEXUALITIES

A number of experts have noted that gay men do not form a homogeneous category, nor do lesbians (e.g., Bell & Weinberg, 1978). This concept is represented by the use of the term *homosexualities* rather than *homosexuality*. Kinsey's 7-point scale assessing various mixtures of heterosexual and homosexual experience in individuals' histories is one recognition of this variability (Kinsey et al., 1948, 1953).

The concept that homosexuals form a heterogeneous category, or that there are multiple subtypes of homosexuals, must be linked to the genetic research. It may be that one subtype of gay men has their orientation because of a gene on the X chromosome, another subtype is the result of polygenic influence, and yet another subtype is the result of prenatal events. Blanchard's group, for example, has found repeatedly that, compared with heterosexual men, gay men are more likely to have a late birth order and to have more older brothers but not older sisters (Blanchard & Bogaert, 1996; Blanchard & Klassen, 1997). They attribute this phenomenon to the formation of H-Y antigen by the mother with each pregnancy with a boy; H-Y antigen is known to influence prenatal sexual differentiation. We look forward to research that identifies similar subtypes and influences among lesbians. Blanchard's research provides one example of new interest among sex researchers in the role of the immune system in sexuality (e.g., Binstock, 2001).

DIFFERENCES BETWEEN GAY MEN AND LESBIANS

Several lines of evidence lead to the conclusion that lesbians differ from gay men on a variety of indicators, suggesting the possibility that genetic influence on sexual orientation is different for men and women. Lesbian women are generally thought to be more bisexual in their behavior and attractions than gay men, who are more likely to be exclusively homosexual. For example, Bailey and colleagues' (Bailey et al., 2000) study of a sample of nonheterosexual women and men examined the distribution of respondents on the Kinsey scale, on which 0 indicates *completely heterosexual feelings* and 6 indicates *completely homosexual feelings*, with the midpoint, 3, indicating *equally heterosexual and homosexual*. Because the sample was nonheterosexual, all respondents had scores in the range from 1 to 6. Women were considerably more likely than men to have a Kinsey score of 1, whereas men were considerably more likely than women to have a Kinsey score of 6. Given that the behavioral and psychological patterns are different for women and men, different explanatory models may be required for the two groups.

Consistent with this observation, some theories—such as Blanchard's theory about birth order and the H-Y antigen predisposing men to homosexuality—address male homosexuality only. For the most part, female homosexuality remains undertheorized.

We should also note, however, that some twin studies show remarkably consistent concordance rates for lesbians compared with gay men—for example, 52% concordance for male MZ twins and 48% concordance for female twins in a pair of studies from the Bailey group (Bailey & Pillard, 1991; Bailey et al., 1993). These findings suggest that homosexual orientation is equally heritable in men and women.

RESEARCH WITH ANIMAL MODELS

Because genetic research on sexual orientation is so difficult with humans, typically is constrained by small sample sizes, and does not permit true experimental designs, some researchers are pursuing animal models for understanding sexual orientation. For example, among domestic sheep 9% of adult males strongly prefer other males as sex partners (Ellis, 1996; Price, Katz, Wallach, & Zenchak, 1988). This line of inquiry immediately raises the question of how one defines sexual orientation in nonhumans. Generally it is defined as animals' attraction to and sexual contact with members of their own sex when members of the other sex are available. The definition, then, is behavioral. We doubt that sheep reflect on their sexual identity.

In *Drosophila* (fruit flies), a number of genes have been identified that control distinct aspects of sexual behavior (Emmons & Lipton, 2003). These include the *period* gene, which controls one aspect of courtship behavior (singing), and the *courtless* gene, which controls one aspect of sex drive. A mutation of the *fruitless* (*fru*) gene leads male fruitflies to have difficulty discriminating between males and females and to mate with both—although this hardly seems comparable to bisexual orientation in humans. One research group has identified mutations of several different genes that result in male–male courtship behavior (Yamamoto & Nakano, 1999).

Zebra finches are a monogamous species that pair-bonds for life. If female finches are exposed to estradiol or fadrozole (an estrogen synthesis inhibitor)—and therefore the developing ova are exposed to these hormones—the genetic female offspring pair with another female in adulthood, even when males are available (Adkins-Regan, 2002). Another experiment by the same group investigated the importance of early rearing environment. Young zebra finches normally grow up in the presence of both a mother and a father, because of the monogamous bond and because both parents participate in the rearing of offspring. If adult males are removed from the cages of their chicks, some of the female offspring in adulthood prefer to bond with another female (Adkins-Regan, 2002). These elegant experiments provide evidence of both biological influence (hormone exposure during early development) and the influence of early rearing environment.

ETHICAL ISSUES

Assume for the moment that in the next 10 years, five distinct genes are identified in humans that are linked to homosexuality, and that if a baby is born carrying any one of these genes, the baby has a 75% chance of being homosexual in adulthood. Assume further that prenatal genetic testing could correctly identify these individuals during the first 3 months of fetal development. At that point, several scenarios are possible. With currently available technology, parents could abort such fetuses if they chose to. This practice currently occurs in many nations around the world in regard to sex selection. Another possibility is that new technology might allow gene substitution so that the gay genes could be replaced by heterosexual genes. When reproduction occurs through assisted reproductive technologies, preimplantation genetic diagnosis would be possible, so that an embryo could be screened for sexual orientation genes and not implanted (discarded) if it were carrying one of the homosexual genes (Dahl, 2003). All of this is within the realm of possibility in the near future and requires serious ethical consideration.

It is beyond the scope of this chapter to explore all of the ethical nuances associated with these issues and the legal strategies that might be necessary depending on the ethical conclusions (for detailed discussions, see Dahl, 2003; Gabard, 1999; Stein, 1998). The arguments, in brief, are as follows. Many parents might feel that homosexuals are the victims of such serious discrimination that the most benevolent approach would be to choose only heterosexual offspring. Others might respond that such an approach advocates abortion for essentially frivolous reasons—that most gays and lesbians lead fulfilling lives and that their condition is not in need of correcting. The dilemma for parents and for society at large would be extraordinary.

CONCLUSIONS

Twin studies with humans provide evidence that homosexual orientation is heritable. At the same time, these studies indicate that sexual orientation is not completely under genetic control, because MZ concordance rates are substantially below 100%. There is some evidence of a single gene on the X chromosome affecting sexual orientation in a subset of homosexual men, although attempts to replicate the findings have produced inconsistent results. Animal models of sexual orientation are being investigated in diverse species, including *Drosophila*, sheep, and finches.

I predict that future research will identify subtypes of gay men and lesbians with different causal pathways to the distinct subtypes or phenotypes; these pathways will include single-gene effects, polygenic influences, and prenatal developmental factors. It is imperative that we as a society consider the ethical issues that will be involved when these biological influences are well identified and parents have the possibility of choosing the sexual orientation of their children.

REFERENCES

Adkins-Regan, E. (2002). Development of sexual partner preference in the zebra finch: A socially monogamous, pair-bonding animal. *Archives of Sexual Behavior, 31*, 27–33.

Bailey, J. M., & Bell, A. P. (1993). Familiality of female and male homosexuality. *Behavior Genetics, 23*, 313–322.

Bailey, J. M., & Benishay, D. S. (1993). Familial aggregation of female sexual orientation. *American Journal of Psychiatry, 150*, 272–277.

Bailey, J. M., Dunne, M. P., & Martin, N. G. (2000). Genetic and environmental influences on sexual orientation and its correlates in an Australian twin sample. *Journal of Personality and Social Psychology, 2000*, 524–536.

Bailey, J. M., & Pillard, R. C. (1991). A genetic study of male sexual orientation. *Archives of General Psychiatry, 48*, 1089–1096.

Bailey, J. M., & Pillard, R. C. (1995). Genetics of human sexual orientation. *Annual Review of Sex Research, 6*, 126–150.

Bailey, J. M., Pillard, R. C., Dawood, K., Miller, M. B. Farrer, L. A., Trivedi, S., & Murphy, R. L. (1999). A family history study of male sexual orientation using three independent samples. *Behavior Genetics, 29,* 79–86.

Bailey, J. M., Pillard, R. C., Neale, M. C., & Agyei, Y. (1993). Heritable factors influence sexual orientation in women. *Archives of General Psychiatry, 50,* 217–223.

Bell, A. P., & Weinberg, M. S. (1978). *Homosexualities.* New York: Simon & Schuster.

Binstock, T. (2001). An immune hypothesis of sexual orientation. *Medical Hypotheses, 57,* 583–590.

Blanchard, R., & Bogaert, A. F. (1996). Homosexuality in men and number of older brothers. *American Journal of Psychiatry, 153,* 27–31.

Blanchard, R., & Klassen, P. (1997). H-Y antigen and homosexuality in men. *Journal of Theoretical Biology, 185,* 373–378.

Dahl, E. (2003). Ethical issues in new uses of preimplantation genetic diagnosis: Should parents be allowed to use preimplantation genetic diagnosis to choose the sexual orientation of their children? *Human Reproductions, 18,* 1368–1369.

Ellis, L. (1996). The role of perinatal factors in determining sexual orientation. In R. C. Savin-Williams & K. M. Cohen (Eds.), *The lives of lesbians, gays, and bisexuals* (pp. 35–70). Fort Worth, TX: Harcourt Brace.

Emmons, S. W., & Lipton, J. (2003). Genetic basis of male sexual behavior. *Journal of Neurobiology, 54,* 93–110.

Gabard, D. L. (1999). Homosexuality and the human genome project: Private and public choices. *Journal of Homosexuality, 37,* 25–51.

Hamer, D., Hu, S., Magnuson, V., Hu, N., & Pattatucci, A. M. (1993, July 16). A linkage between DNA markers on the X chromosome and male sexual orientation. *Science, 261,* 321–327.

Hershberger, S. L. (1997). A twin registry study of male and female sexual orientation. *Journal of Sex Research, 34,* 212–222.

Heston, L., & Shields, J. (1968). Homosexuality in twins: A family study and a registry study. *Archives of General Psychiatry, 18,* 149–160.

Hu, S., Pattatucci, A., Patterson, C., Li, L., Fulker, D., Cherny, S., et al. (1995). Linkage between sexual orientation and chromosome Xq28 in males but not females. *Nature Genetics, 11,* 248–256.

Hyde, J. S., & DeLamater, J. D. (2003). *Understanding human sexuality* (8th ed.). New York: McGraw-Hill.

Kallmann, F. J. (1952a). Comparative twin study on the genetic aspects of male homosexuality. *Journal of Nervous and Mental Disease, 115,* 283–298.

Kallmann, F. J. (1952b). Twin and sibship study of overt male homosexuality. *American Journal of Human Genetics, 4,* 136–146.

Kendler, K. S., Thornton, L. M., Gilman, S. E., & Kessler, R. C. (2000). Sexual orientation in a U.S. national sample of twin and nontwin sibling pairs. *American Journal of Psychiatry, 157,* 1843–1846.

King, M. N., & McDonald, E. (1992). Homosexuals who are twins: A study of 46 probands. *British Journal of Psychiatry, 160,* 407–409.

Kinsey, A. C., Pomeroy, W. B., & Martin, C. E. (1948). *Sexual behavior in the human male.* Philadelphia: Saunders.

Kinsey, A. C., Pomeroy, W. B., Martin, C. E., & Gebhard, P. H. (1953). *Sexual behavior in the human female.* Philadelphia: Saunders.

Laumann, E. O., Gagnon, J. H., Michael, R. T., & Michaels, S. (1994). *The social organization of sexuality: Sexual practices in the United States.* Chicago: University of Chicago Press.

Lever, J., Kanouse, D. E., & Rogers, W. H. (1992). Behavior patterns and sexual identity of bisexual males. *Journal of Sex Research, 29,* 141–167.

Marshall, E. (1999, October 15). An array of uses: Expression patterns in strawberries, Ebola, TB, and mouse cells. *Science, 286,* 445.

Miki, Y., Swensen, J., Shattuck-Eidens, D., Futreal, P. A., Harshman, K., Tavtigian, S., et al. (1994, October 7). A strong candidate for the breast and ovarian cancer susceptibility gene BRCA1. *Science, 226,* 66–71.

Mustanski, B. S., Chivers, M. L., & Bailey, J. M. (2002). A critical review of recent biological research on human sexual orientation. *Annual Review of Sex Research, 13,* 89–140.

Mustanski, B. S., DuPree, M. G., Nievergelt, C. M., Bocklandt, S., Schork, N. J., & Hamer, D. (2005, March). A genomewide scan of male sexual orientation. *Human Genetics.*

Pillard, R. C., & Bailey, J. N. (1998). Human sexual orientation has a heritable component. *Human Biology, 70,* 347–365.

Price, E. O., Katz, L. S., Wallach, S. J. R., & Zenchak, J. J. (1988). The relationship of male–male mounting to the sexual preferences of young rams. *Applied Animal Behaviour Science, 21,* 347–355.

Rice, G., Anderson, C., Risch, N., & Ebers, G. (1999, April 23). Male homosexuality: Absence of linkage to microsatellite markers at Xq28. *Science, 284,* 665–667.

Stein, E. (1998). Choosing the sexual orientation of children. *Bioethics, 12,* 1–24.

Whitam, F. L., Diamond, M., & Martin, J. (1993). Homosexual orientation in twins: A report on 61 pairs and three triplet sets. *Archives of Sexual Behavior, 22,* 187–206.

Yamamoto, D., & Nakano, Y. (1999). Sexual behavior mutants revisited: Molecular and cellular basis of Drosophila mating. *Cellular and Molecular Life Sciences, 56,* 634–646.

3

THE ROLE OF HYPOTHALAMUS AND ENDOCRINE SYSTEM IN SEXUALITY

DICK F. SWAAB

Recent research reveals that the hypothalamus and its endocrine systems exert strong control over sexual behavior and sexual differences. In this chapter, I explore the nature and workings of this control mechanism. First, I discuss the mechanisms of sexual differentiation of the brain in humans and animals, and then I explore the particular way that the hypothalamus is differentiated and how different hormone levels affect this process. Finally, I discuss the control of these systems over sexual behavior and their possible effect on gender and sexual orientation.

Sexual difference in brain structures necessarily begins to develop early in any organism. In this section, I explain the various mechanisms of sexual differentiation. A glossary is included at the end of this chapter.

MECHANISM OF SEXUAL DIFFERENTIATION OF THE BRAIN: THE AROMATIZATION THEORY

Testosterone is formed in the fetal Leydig cells in the testicle from about 8 weeks of gestation onward (Hiort, 2000). Sexual differentiation of

I am grateful to W. T. P. Verweij for her excellent secretarial and linguistic help, to J. Kruisbrink for bibliographic assistance, and to G. van der Meulen and H. Stoffels for the illustrations.

21

the brain is thought to be imprinted or organized by hormonal signals from the developing male gonads. On the basis of animal experiments, this process is presumed to be induced by androgens during development, following this local conversion in the brain to estrogens by P-450 aromatase. In analogy, male sexual differentiation of the human brain is thus most probably determined in the two first periods during which peaks in testosterone levels are found in boys and not in girls, namely (a) during gestation and (b) during the perinatal period. Subsequently, (c) from puberty onward, sex hormones alter the function of previously organized neuronal systems ("activating effects") in both sexes (for references, see De Zegher, Devlieger, & Veldhuis, 1992; Forest, 1989; Swaab, Gooren, & Hofman, 1992; Tucker Halpern, Udry, & Suchindran, 1998). In late gestation, the male fetus is exposed to both high concentrations of testosterone from itself and high levels of estrogens from the placenta. After the inhibitory effect of maternal estrogens wanes, postnatally, the infant gonadotropins increase, resulting in a rapid rise of testosterone until some 6 months after birth. The female fetus is exposed only to estrogens from the placenta. In postnatal female infants the surge of follicle-stimulating hormone predominates, giving rise to an increase in estradiol 2 to 4 months postnatally (Quigley, 2002). The possible importance of the male testosterone surge for sexual differentiation of the brain following aromatization to estrogens agrees with the observation that girls whose mothers were exposed to diethylstilboestrol (DES) during pregnancy have an increased chance of occurrence of bi- or homosexuality (Ehrhardt et al., 1985; Meyer-Bahlburg et al., 1995; see Table 3.1). Estrogens produced locally by aromatase are thought to participate in numerous biological functions, including sexual differentiation of the brain. Aromatase is found throughout the adult human brain, with the highest levels in the pons, thalamus, hypothalamus, and hippocampus. The amount of aromatase mRNA is highest in the hypothalamus, thalamus, and amygdala. No differences were detected between the 4 men and 2 women studied (Sasano, Takahasi, Satoh, Nagura, & Harada, 1998). Such a study has, however, so far not been performed in larger series and also in development. The neonatal peak of testosterone is probably not induced by the hypothalamus of the child but rather by chorionic gonadotrophins from the placenta, because a normal neonatal increase of testosterone was found in a patient with hypogonadism based on a DAX-1 gene mutation that causes hypogonadotrophic hypogonadism (Takahashi et al., 1997).

Although both sexes are exposed to high placental estrogen levels prenatally, only males are subjected to high levels of androgens (Quigley, 2002). In human neonates, at 34 to 41 weeks of gestation the testosterone level is 10-fold higher in men than in women (De Zegher et al., 1992). Although the peak in serum testosterone in boys at 1–3 months postnatally approaches the levels seen in adult men, most testosterone is bound to

TABLE 3.1
Factors That Influence Sexual Differentiation of the Human Brain

Gender Identity (transsexualism)

Chromosomal disorders	Rare: 47 XYY (male-to-female), 47 XXX (female-to-male; Hengstschläger et al., 2003; Turan et al., 2000).
	Microdeletion on Y-chromosome in 1 male-to-female transsexual.
	Steroidogenic factor-1 mutations (give XY sex-reversal, no transsexuality; Achermann, Meeks, & Jameson, 2001; Achermann, Weiss, Lee, & Jameson, 2001; Correa et al., 2004; Ozisik, Achermann, & Jameson, 2002).
	Klinefelter XXY male-to-female (Sadeghi & Fakhrai, 2000).
	Twin studies (Coolidge et al., 2002; Sadeghi & Fakhrai, 2000).
	Genomic imprinting (Green & Keverne, 2000).
Phenobarbital/diphantoin	(Dessens et al., 1999)
Hormones	Intersex (Reiner, 1996; Zucker et al., 1987), micropenis (Reiner, 2002).
	Cloacal exstrophy (Reiner & Gearhart, 2004; Zderic et al., 2002; Zucker, 2002).
	5 α-reductase deficiency, 17β-hydroxy-steroid-dehydrogenase-3 deficiency (Imperato-McGinley et al., 1979, 1991; Wilson, 1999).
	Girls with congenital adrenal hyperplasia (CAH) with gender problems (Meyer-Bahlburg et al., 1996; Slijper et al., 1998; Zucker et al., 1996).
	More polycystic ovaries, oligomenorrhea, and amenorhea are found in transsexuals (Futterweit et al., 1986).
	Complete androgen insensitivity syndrome results in XY heterosexual females (Wisniewski et al., 2000).
Social factors?	(Bradley et al., 1998) not effective: John/Joan/John case (Cohen-Kettenis & Gooren, 1998; Diamond & Sigmundson, 1997).

Sexual Orientation (homosexuality, heterosexuality)

Genetic factors	Twin studies (Bailey & Bell, 1993; Kallman, 1952).
	Molecular genetics (Hamer et al., 1993; Hu et al., 1995); however, see Rice et al. (1999).
Hormones	Girls with CAH (Dittmann et al., 1992; Money et al., 1984; Zucker et al., 1996).
	DES (Ehrhardt et al., 1985; Meyer-Bahlburg et al., 1995).
	Male-to-female sex reassignment (Bailey et al., 1999).
Chemicals	Nicotine prenatally increases the occurrence of lesbianism (Ellis & Cole-Harding, 2001).
Social factors?	Stress during pregnancy (Bailey et al., 1991; Ellis et al., 1988; Ellis & Cole-Harding, 2001).
	Raising by transsexual or homosexual parents does not affect sexual orientation (Golombok et al., 1983; Green, 1978).

globulin. Yet the amount of free testosterone in male infants is about one order of magnitude larger than that in female infants at this time (Bolton, Tapanainen, Koivisto, & Vihko, 1989). During the adrenarche (i.e., from 7 years of age to the onset of puberty), the adrenal starts to produce more androgens, predominantly DHEA (dehydroepiandrosterone) and DHEAS (dehydroepiandrosterone sulphate). After the age of 8, testosterone from the adrenal also starts to rise, while a small but significant testicular production of testosterone is also present in prepubertal boys (Forest, 1989). Although the testosterone peak during puberty (Forest, 1989) is generally thought to be involved in activation rather than organization, the neuron number of the female domestic pig hypothalamus showed a surprising twofold increase in a sexually dimorphic hypothalamic nucleus around puberty (Van Eerdenburg & Swaab, 1991), which means that, although this phenomenon may have been programmed earlier, late organizational effects can at present not be excluded. Few data are available on the exact period in development when the human brain differentiates according to sex. Brain weight is sexually dimorphic from 2 years postnatally onward, taking differences in body weight between boys and girls into account (Swaab & Hofman, 1984). As I discuss later, sexual differentiation of the human *sexually dimorphic nucleus of the preoptic area* (SDN-POA) becomes apparent between 4 years and puberty (Swaab & Hofman, 1988), and a similar late sexual differentiation was found in the darkly staining posteromedial component of the *bed nucleus of the stria terminalis* (BST; Allen, Hines, Shryne, & Gorski, 1989a). In the *central nucleus of the BST* (BSTc), the sex difference did not become significant until early adulthood (Chung, De Vries, & Swaab, 2002). One might conclude that the limited evidence that is currently available suggests that sexual differentiation of the human hypothalamus becomes apparent between 2 years of age and young adulthood, although this may, of course, be based on processes that were programmed much earlier, for instance by a peak in sex hormone levels in mid-pregnancy or during the neonatal period (see before). On the basis of existing prenatal serum samples from the mother, it appeared that, indeed, higher androgen exposure in the second trimester of fetal life may masculinize a girl's behavior (Udry, Morris, & Kovenock, 1995).

Direct Effect of Testosterone

Although the process by which estrogens are derived from testosterone by aromatization is considered to be the major mechanism for androgenization of the brain during development in rodents, testosterone itself may be of major importance for sexual differentiation of the human brain. In the first place, both sexes are exposed to high levels of estrogens during fetal life, whereas only males are subjected to high androgen levels (Quigley, 2002). Moreover, the androgen receptor, from which the gene that is located on the X-chromosome at Xq11–12, may be mutated in such a way that the subject has a complete androgen insensitivity syndrome. Despite normal testis differentiation and androgen biosynthesis, the phenotype has a normal female external and behavioral appearance (Batch et al., 1992). Phenotypic women with complete androgen insensitivity syndrome perceive themselves as highly feminine. They do not have gender problems and largely report their sexual attraction, fantasies, and experiences as heterosexual women (Wilson, 1999; Wisniewski et al., 2000). This means that for the development of human male gender identity and male heterosexuality, direct androgen action on the brain seems to be of crucial importance, and that the aromatization theory may even be of secondary importance for human sexual differentiation of the brain. The observation by Macke et al. (1993) that DNA sequence variation in the androgen receptor is not a common determinant of sexual orientation seems to be at variance with these data, but it should be noted that the DNA variations in that study did not prevent normal androgenization of the subjects studied, so there was no loss of function of this receptor. Experiments in rodents and nonhuman primates now also indicate the importance of androgens for the masculinization of the brain, not only by regulating transcription of aromatase mRNA (Roselli, Abdelgadir, & Resko, 1997; Roselli & Resko, 2001) but also by a direct effect on the androgen receptor (Sato et al., 2004; Roselli & Resko, 2001). This point of view agrees with the lack of gender problems in a brother and a sister with aromatase deficiency due to a mutation (Morishima, Grumbach, Simpson, Fisher, & Qin, 1995). Moreover, a 28-year-old man with estrogen resistance due to a mutation of the estrogen-receptor gene was described as tall, with continued linear growth in adulthood, incomplete epiphysal closure, and increased estrogen and gonadotrophin levels. A change in a single base pair in the second exon of the estrogen-receptor gene was found. However, the patient did not report a history of gender-identity disorder, had a strong heterosexual interest, and had normal male genitalia. The elevated serum estrogen levels are explained by a possible compensatory increase in aromatase activity in response to estrogen resistance (Smith et al., 1994). The observations in complete androgen sensitivity syndrome (Wisniewski et al., 2000) and the male gender heterosexual behavior of patients with 5 α-reductase-2 or 17β-hydroxy-steroid dehydrogenase-3 deficiency (Imperato-McGinley, Peterson, Gautier, & Sturla, 1979; Imperato-McGinley et al., 1991; Wilson, 1999; Wilson, Griffin, & Russell, 1993) indicate that a direct action of testosterone may be more important than one of dihydrotestosterone or testosterone aromatized to estrogens for male heterosexual psychosexual development.

Neurotransmitters and Genes

Not only may sex hormones affect sexual differentiation of the brain, on the basis of animal experiments it is expected that all compounds that influence hormone or neurotransmitter metabolism in development may also affect sexual differentiation of the brain (Pilgrim & Reisert, 1992). For example, young adult male mice that were prenatally exposed to alcohol were found to have a decreased preference for female partners and an increased preference for males (Watabe & Endo, 1994). Exposure during development to some drugs (e.g., barbiturates) causes deviations in testosterone levels, persisting into adulthood (Ward, 1992). This agrees with the finding of Dessens et al. (1999) that children born of mothers who were exposed to anticonvulsants such as phenobarbital and diphantoin have an increased probability of transsexuality (see below). Exposure of rats to drugs such as opiates led to behavioral changes despite apparently normal adult gonadal hormone levels (Ward, 1992). Similar observations in human sexual differentiation have not yet been reported. It is of great interest that, in addition, there is animal experimental evidence for primary genetic control of sexual differentiation that does not involve sex hormones. Results obtained from cultures of embryonic rat brain indicate that dopaminergic neurons may develop morphological and functional sex differences in the absence of sex steroids (Pilgrim & Reisert, 1992). Candidates for such hormone-independent effects are those genes located on the nonrecombining part of the Y-chromosome and believed to be involved in primary sex determination of the organism. Two candidate genes are the two testis-determining factors ZFY and the master switch for differentiation of a testis SRY. Those are putative transcription factors. It has been shown that SRY and ZFY are transcribed in the hypothalamus and frontal and temporal cortex of adult men and not in women. It may well be possible that they function as sex-specific cell-intrinsic signals that are needed for full differentiation of a male human brain, and that continuous expression throughout life may be required to maintain sex-specific structural or functional properties of differentiated male neurons. Sexual differentiation of the human brain may thus be a multifactorial process, although a role of SRY and ZFY in this process still needs to be proved (Mayer, Lahr, Swaab, Pilgrim, & Reisert, 1998). An alternative mechanism could be the actions of an imprinted X-linked locus (Skuse, 1999). Microarray studies in mice have shown that there are over 50 genes expressed in a sexually dimorphic way in the brain before gonadal differentiation takes place (Dewing, Shi, Horvath, & Vilain, 2003). The relative contributions of the different sex hormones and other nonhormonal factors on sexual differentiation of the human brain should clearly be a focus for future research.

Sexual Differentiation, the Hypothalamus, and the Amygdala

Sex differences in the hypothalamus and other limbic structures are thought to be the basis of sex differences in sexual arousal (Karama et al., 2002), reproduction (e.g., copulatory behavior in both sexes, the menstrual cycle in women), gender identity (i.e., the feeling one is either a man or a woman), gender identity disorders (transsexuality), and sexual orientation (homosexuality, heterosexuality, bisexuality; Gooren, Fliers, & Courtney, 1990; Swaab et al., 1992; Swaab & Hofman, 1995). In addition, disinhibited sexual activity and paraphilias have been reported following lesions in the hypothalamus and septum (Frohman, Frohman, & Moreault, 2002; Miller, Cummings, McIntyre, Ebers, & Grode, 1986). The *paraventricular nucleus* (PVN) in rat, and in particular its oxytocin neurons, are involved in erectile functions *in copula*. Also, in monkey, penile erection is induced by electrical stimulation of the PVN (MacLean & Ploog, 1962). Not only oxygen but also α-MSH induces erection in rats after intracerebroventricular injection (Mizusawa, Hedlund, & Andersson, 2002). In humans, the melanocortin receptor agonist MTII was proerectile in men with organic and psychogenic erectile dysfunction (MacNeil et al., 2002; Van der Ploeg et al., 2002). PT-141, another melanocortin agonist, also appeared to be erectogenic in normal men and in patients with erectile dysfunction. In rodents it appeared to act by way of the PVN (Molinoff, Shadiack, Earle, Diamond, & Quon, 2003). For penile erections initiated by psychogenic stimuli, the medial amygdala, the PVN, and to a lesser extent the BST are involved, whereas medial preoptic lesions in the rat have little, if any, effect on this type of erection (Liu, Salamone, & Sachs, 1997). Sexual arousal and orgasm produce long-lasting alterations in plasma prolactin concentrations, both in men and women (Exton et al., 1999), which might be related to postorgasmic loss of arousability (Bancroft, 1999), indicating a role of the medial-basal hypothalamic dopamine neurons.

There is extensive animal experimental literature that shows that the medial preoptic area (mPOA) of the hypothalamus is a key structure for male and female copulatory behavior (McKenna, 1998; Pilgrim & Reisert, 1992; Yahr, Finn, Hoffman, & Sayag, 1994) and that the hypothalamus is involved in seminal vesicle contractions at coitus (Cross & Glover, 1958). Electrical stimulation of the preoptic area in monkey induces penile erection (MacLean & Ploog, 1962). However, the exact role of its subarea, the SDN-POA, in these functions is not clear, and literature on hypothalamic structures involved in sexual orientation in experimental animals is scarce (Kindon, Baum, & Paredes, 1996; Paredes, Tzschentke, & Nakach, 1998). Paredes and Baum (1995) found that lesions of the mPOA/anterior hypothalamus in the male ferret not only affected masculine coital performance but also affected heterosexual partner preference. Perkins and colleagues showed that testosterone, estron, and estradiol plasma concentrations were higher in female-oriented rams than in homosexual rams. In the preoptic area, the aromatase activity was higher in the female-oriented than in the male-oriented rams, indicating again the possible importance of the preoptic area in sexual orientation (Resko et al., 1996). Edwards, Walter, and Liang (1996) observed a decrease in partner preference in male rats following a lesion of the BST. The content of estrogen receptors in the amygdala was found to be similar in both homosexual rams and ewes but less than the receptor content in heterosexual rams, whereas the estrogen receptor content of the hypothalamus, anterior pituitary, and preoptic area of these two groups did not differ. These data, and the observations in Klüver-Bucy syndrome (see below), suggest that the amygdala might also play a role in sexual orientation. In sheep, an ovine sexually dimorphic nucleus was identified in the preoptic area of the hypothalamus that was two times larger in female-oriented rams than in male-oriented ones. The aromatase mRNA levels in this nucleus were greater in female-oriented rams than in ewes, whereas the male-oriented rams had intermediate levels (Roselli, Larkin, Resko, Stellflug, & Stormshak, 2004).

There is no information available on the exact nature of the preoptic and amygdala neuronal systems and connections that may be involved in sexual orientation. Data on the hypothalamus in relation to gender identity in animals are, of course, nonexistent.

Observations in Humans

A few studies have been presented in the medical literature that implicate the hypothalamus and adjoining structures in various aspects of sexual behavior of human and nonhuman primates. In some Klüver-Bucy syndrome cases, because of damage of the temporal lobe, patients were reported to change from heterosexual to homosexual behaviors, indicating that the temporal lobe might be of importance for sexual orientation (Lilly, Cummings, Benson, & Frankel, 1983; Marlowe, Mancall, & Thomas, 1975; Terzian & Dalle Ore, 1955). Also, there are a few case histories of changing sexual orientation, from heterosexual to pedophilic or homosexual, on the basis of a lesion in the hypothalamus or in the temporal lobe, from which the amygdala has strong connections to the hypothalamus (Miller et al., 1986). Direct electrical or chemical stimulation of the septum may induce a sexually motivated state of varying degrees up to penile erection in men and building up to an orgasm in both sexes (Heath, 1964). Markedly increased sexual behavior was observed following the placement of the tip of a ventriculoperitoneal shunt into the septum in two cases (Gorman & Cummings, 1992). Meyers (1961) described a loss of potency following lesion in the septo-fornico-hypothalamic region. Electrical stimulation of the mamillary body in monkeys induced penile erection (MacLean & Ploog, 1962; Poeck & Pilleri, 1965). Precocious puberty and hypersexuality have been reported following lesions in the posterior part of the hypothalamus, and hypogonadism is an early sign of pathology in the anterior part of the hypothalamus (Bauer, 1954, 1959; Poeck & Pilleri, 1965). In particular, hamartoma that affect the posterior region of the hypothalamus, that is, those that are pendulated and attached to the region of the mamillary bodies, may cause precocious puberty (Valdueza et al., 1994). In addition, precocious puberty is found in cases of pineal region tumors that produce gonadotropins.

A German stereotactic psychosurgical study (Müller, Roeder, & Orthner, 1973) reported on 22 mainly pedo- or ephebophilic (i.e., preferring pubertal boys) homosexual men (N = 14) and 6 cases with disturbances of heterosexual behavior (hypersexuality, exhibitionism, or pedophilia). In 12 homosexual patients, a lesion was made in the right ventromedial nucleus of the hypothalamus. In 8 patients, homosexual fantasies and impulses disappeared. According to Müller et al., in 6 patients a "vivid desire for full heterosexual contacts" occurred after the operation. In 1 pedophilic patient, bilateral destruction of the ventromedial nucleus was performed and he lost all interest in sexual activity after the operation. The heterosexual patients reported a significant reduction of their sexual drive. Unilateral ventromedial hypothalamotomy in 14 cases treated for aggressive sexual delinquency caused a decrease in sexual drive (Albert, Walsh, & Jonk, 1993; Dieckmann, Schneider-Jonietz, & Schneider, 1988). Although these studies at first sight appear to suggest that the human ventromedial nucleus of the hypothalamus is indeed involved in sexual orientation and sexual drive, they are highly controversial from an ethical point of view and methodologically deficient (Heimann, 1979; Rieber & Sigusch, 1979; Schorsch & Schmidt, 1979).

SEX DIFFERENCES IN THE HYPOTHALAMUS

Sex differences in brain and hormone levels not only are of importance for sexual behavior but are also thought to be the structural and functional basis of sex differences in cognition and other central functions, and for the often pronounced sex differences in the prevalence of neurological and psychiatric diseases (Swaab, 2002). There is an increasing amount of data concerning morphological and functional sex differences, especially in the various nuclei of the hypothalamus and adjacent structures that are so important for sexual behavior (see Figures 3.1, 3.2, and 3.3). Functional magnetic resonance imaging (fMRI) revealed a significant activation of the right-hand-side hypothalamus in males who were sexually aroused (Arnow et al., 2002).

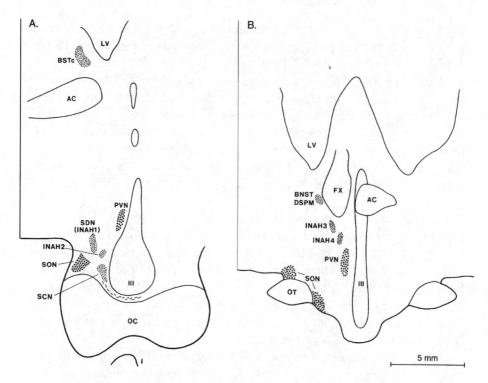

Figure 3.1. Topography of the sexually dimorphic structures in the human hypothalamus. A (left) is a more rostral view than B (right). Abbreviations: III = third ventricle; AC = anterior commissure; BNST-DSPM = darkly staining posteriomedial component of the bed nucleus of the stria terminalis; BSTc = central nucleus of the bed nucleus of the stria terminalis; FX = fornix; I = infundibulum; INAH1–4 = interstitial nucleus of the anterior hypothalamus 1–4; LV = lateral ventricle; OC = optic chiasm; OT = optic tract; PVN = paraventricular nucleus; SCN = suprachiasmatic nucleus; SDN = sexually dimorphic nucleus of the preoptic area; SON = supraoptic nucleus. Scale bar = 5 mm. The AC, BSTc, BNST-DSPM, INAH2,3,4, SCN, and SDN vary according to sex. The SCN, INAH3, and AC are different in relation to sexual orientation.

The hypothalamus contains a number of structurally sexually dimorphic structures (Figures 3.1, 3.2, and 3.3) that are presumed to be involved in sexual behavior, such as the SDN-POA, which was first described in the rat brain by Gorski, Gordon, Shryne, and Southam (1978). *Morphometric analysis* revealed that there is also a striking sexual dimorphism in the size and cell number in the human SDN-POA (Hofman & Swaab, 1989; Swaab & Fliers, 1985; Swaab & Hofman, 1984; Figures 3.3 and 3.4). In terms of cell numbers, sexual differentiation of the human SDN-POA occurs after 4 years postnatally (Figure 3.5; see below). This is due to a decrease in both volume and cell number in women, whereas in men volume and cell number remain unaltered up to the fifth decade, after which a marked decrease in cell number is observed as well (Figure 3.4; see below). Sex differences

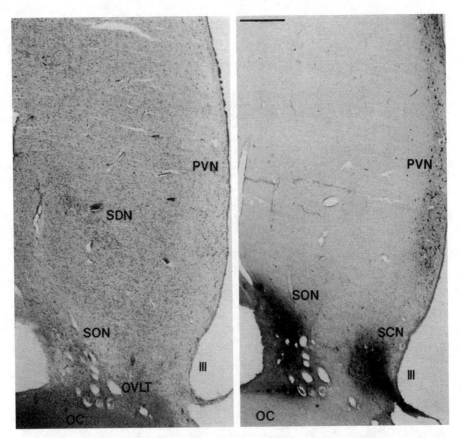

Figure 3.2. Thionine (left) and antivasopressin (right) stained section through the chiasmatic or preoptic region of the hypothalamus. OC = optic chiasm; OVLT = organum vasculosum lamina terminalis; PVN = paraventricular nucleus; SCN = suprachiasmatic nucleus; SDN = sexually dimorphic nucleus of the preoptic area (intermediate nucleus, INAH-1); SON = supraoptic nucleus; III = third ventricle. Bar represents 1 mm.

have also been reported for two other cell groups in the preoptic-anterior-hypothalamus (INAH 2 and 3; Allen, Hines, Shryne, & Gorski, 1989b). These areas (Figure 3.1) were found to be larger in men than in women, which was later partly confirmed by LeVay (1991) and Byne et al. (2000). Other brain regions with a larger volume in men than in women are the darkly staining posteromedial part of the bed nucleus of the stria terminalis (BST-dspm), as described by Allen et al. (1989a), and the BSTc (Zhou, Hofman, Gooren, & Swaab, 1995; Figure 3.6). The sex difference in the number of vasoactive intestinal polypeptide expressing neurons in the *suprachiasmatic nucleus* (SCN) is age-dependent (Zhou, Hofman, & Swaab, 1995b). The *infundibular nucleus* shows sex-dependent Alzheimer changes (Schultz, Braak, & Braak, 1996), and the size of the *anterior commissure* is

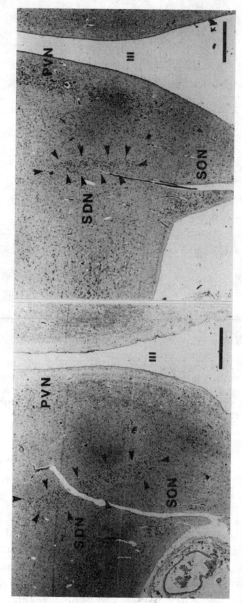

Figure 3.3. Thionin-stained frontal section (6:m) of the hypothalamus of (left) a 28-year-old man and (right) a 10-year-old girl. Arrows show the extent of the sexually dimorphic nucleus of the preoptic area (SDN). Note the large blood vessel penetrating the SDN-POA and note that the SDN of the man is larger than that of the girl. PVN = paraventricular nucleus; SON = supraoptic nucleus; III = third ventricle. Bar represents 1 mm. From "A Sexually Dimorphic Nucleus in the Human Brain," by D. F. Swaab and E. Fliers, 1985, *Science, 228*, p. 1113. Copyright 1985 by the American Association for the Advancement of Science. Reprinted with permission.

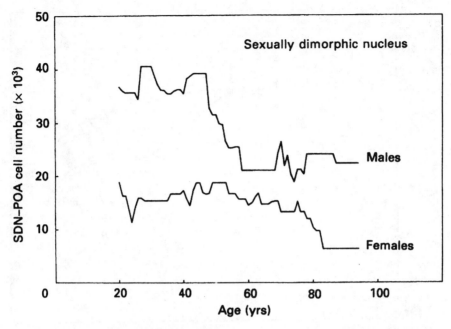

Figure 3.4. Age-related changes in the total cell number of the sexually dimorphic nucleus of the preoptic area (SDN-POA) in the human hypothalamus. The general trend in the data is enhanced by using smoothed growth curves. Note that in males SDN-POA cell number steeply declines between the ages of 50 and 70 years, whereas in females a more gradual cell loss is observed around the age of 70 years. These curves demonstrate that the reduction in cell number in the human SDN-POA in the course of aging is a nonlinear, sex-dependent process. From "The Sexually Dimorphic Nucleus of the Preoptic Area in the Human Brain: A Comparative Morphometric Study," by M. A. Hofman and D. F. Swaab, 1989, *Journal of Anatomy, 164,* p. 63. Copyright 1989 by the Anatomical Society of Great Britain and Ireland. Reprinted with permission.

sexually dimorphic (Allen & Gorski, 1991). 5-hydroxyindoleacetic acid levels in the hypothalamus of men were lower than those in women, indicating a higher turnover rate of 5HT (serotonin) in the female brain (Gottfries, Roos, & Winblad, 1974). In addition, we found clear age-dependent sex differences in the activity of the vasopressinergic neurons of the *supraoptic nucleus* (SON; Ishunina, Kruijver, Balesar, & Swaab, 2000). A few of these sex differences are discussed in more detail below.

Sexually Dimorphic Nucleus of the Preoptic Area

There are indications that the area in which the SDN-POA (*interstitial nucleus of the anterior hypothalamus*; INAH-1) is situated is involved in sexual behavior. Electrical stimulation of the median preoptic area in the rat induces highly exaggerated stimulation-bound sexual behavior (Merari

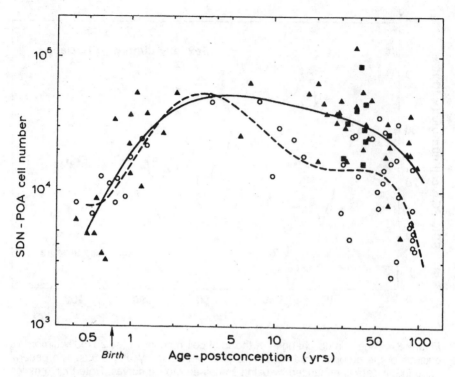

Figure 3.5. Developmental and sexual differentiation of the human sexually dimorphic nucleus of the preoptic area (SDN-POA) of the hypothalamus in 99 subjects, log-log scale. Note that at the moment of birth the SDN-POA is equally small in boys (▲) and girls (○) and contains about 20% of the cell number found at 2 to 4 years of age. Cell numbers reach a peak value around 2 to 4 years postnatally, after which a sexual differentiation occurs in the SDN-POA due to a decrease in cell number in the SDN-POA of females, whereas the cell number in males remains approximately unchanged up to the age of 50. The SDN-POA cell number in homosexual men (■) does not differ from that in the male reference group. The curves are quintic polynomial functions fitted to the original data for men (full line) and women (dashed line). From "Sexual Differentiation of the Human Hypothalamus: Ontogeny of the Sexually Dimorphic Nucleus of the Preoptic Area," by D. F. Swaab and M. A. Hofman, 1988, *Developmental Brain Research, 44,* p. 316. Copyright 1988 by Elsevier Science. Adapted with permission.

& Ginton, 1975). Electrical stimulation of the preoptic area elicited penile erection in monkeys (MacLean & Ploog, 1962). Although following lesions of the mPOA, mounting, intromission, and ejaculation are eliminated, the animals do not lose the ability to achieve an erection (McKenna, 1998). Although the mPOA does not seem to organize copulatory behavior, it is crucial for the recognition of sensory stimuli as appropriate sexual targets and for the integration of this recognition with sexual motivation and copulatory motor programs. The PVN receives extensive input from the mPOA of the preoptic area. The *oxytocinergic neurons* of the PVN project to

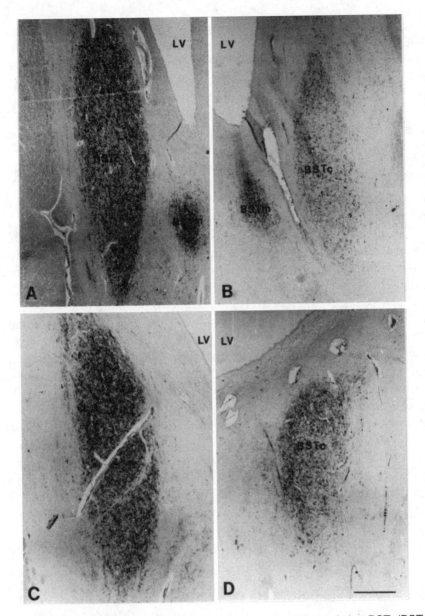

Figure 3.6. Representative sections of the central nucleus of the BST (BSTc) innervated by vasoactive intestinal polypeptide (VIP): (A) heterosexual man; (B) heterosexual woman; (C) homosexual man; (D) male-to-female transsexual. Scale bar = 0.5 mm. BST = bed nucleus of the stria terminalis; LV = lateral ventricle. Note there are two parts of the BST in A and B: small medial subdivision (BSTm) and large oval-sized central subdivision (BSTc). A female volume was observed in genetically male (male-to-female) transsexuals. There was no difference according to sexual orientation. From "A Sex Difference in the Human Brain and Its Relation to Transexuality," by J. N. Zhou, M. A. Hofman, L. J. G. Gooren, and D. F. Swaab, 1995, *Nature, 378,* pp. 68–70. Copyright 1995 by Nature Publishing Group. Reprinted with permission.

the spinal cord and synapse on neurons that innervate the penis (McKenna, 1998). Electrophysiological experiments in rats showed increased multiple unit activity in the preoptic area during mounting, both in intact males and females (Kartha & Ramakrishna, 1996). However, the exact role of the SDN-POA in these functions of the preoptic area is unclear. Although lesion experiments in rats did indicate that the SDN-POA may be involved in aspects of male sexual behavior, that is, mounting, intromission, and ejaculation (De Jonge et al., 1989; Turkenburg, Swaab, Endert, Louwerse, & Van de Poll, 1988), the effects of lesions on sexual behavior were only slight, so that it may well be that the major functions of the SDN-POA are still unknown. Penile erection following mPOA stimulation in monkey and rat as reported by Giuliano et al. (1996) is considered one of such putative functions by some authors, but following lesions of the mPOA, rats do not lose the ability to achieve an erection (McKenna, 1998). Another possible function as a *galanin*-containing sleep-regulating area, homologous to the ventrolateral preoptic nucleus in the rat (Gaus, Strecker, Tate, Parker, & Saper, 2002), is not very probable because of the more medial location of the SDN-POA in the human. On the basis of a number of animal experiments, the SDN-POA area is presumed to be involved in sexual orientation. Lesions of the area of the SDN-POA in the ferret caused a significant shift in the males' preference from estrous females to stud males, that is, from a male-typical pattern of sexual behavior to a more female-typical pattern (Kindon et al., 1996; Paredes & Baum, 1995). Intact female rats show a preference for interaction with males, and males show a tendency to interact with females. After lesion of the mPOA, the females' preference was not modified. However, mPOA-lesioned male rats changed their partner preference after the lesion, and the coital behavior of these males was significantly reduced after the lesion (Paredes et al., 1998). Although in our studies of the human SDN-POA we did not find differences in size or cell numbers of this nucleus in relation to sexual orientation (Swaab & Hofman, 1990), this does, of course, not exclude a functional involvement of this structure in sexual orientation.

Homology to the Rat SDN-POA

The SDN-POA in the young adult human brain is twice as large in men (0.20 mm^3) as in women (0.10 mm^3 on one side) and contains twice as many cells (Swaab & Fliers, 1985; Figures 3.3 and 3.4). The original observations on 13 men and 18 women were extended and confirmed in a group of 103 subjects containing a reference group of 42 men and 38 women who did not die of a primary neurological or psychiatric disease (Swaab & Hofman, 1988). Its sexual dimorphism in the human has been confirmed by H. Braak (see Braak & Braak, 1992, p.14). The SDN-POA is also present in rhesus monkey (Braak & Braak, 1992). The SDN-POA is identical to the intermediate nucleus described by Brockhaus (1942) and Braak and Braak (1987) and to the interstitial nucleus of the anterior hypothalamus-1 (INAH-1) of Allen et al. (1989b). Judging by the sex difference in the human SDN-POA in size and cell number, rostrocaudal position, cytoarchitecture, peptide, and *gamma-amino butyric acid* (GABA) content (see below), this nucleus is most probably homologous to the SDN-POA in the rat of Gorski et al. (1978), despite the fact that the rat SDN-POA is located in a more medial position than its human counterpart (Koutcherov, Mai, Ashwell, & Paxinos, 2002). The SDN-POA contains galanin (Gai, Geffen, & Blessing 1990), named after its N-terminal residue glycine and its c-terminal alanine (Tatemoto, Rökaeus, Jörnvall, McDonald, & Mutt, 1983), and galanin-mRNA (Bonnefond, Palacios, Probst, & Mengod, 1990; Gaus et al., 2002). It should be noted that microinjection of galanin in the medial preoptic nucleus facilitates female-typical and male-typical sexual behaviors in the female rat (Bloch, Butler, & Kohlert, 1996). In addition, the SDN-POA contains *thyrotropin-releasing hormone* (TRH) neurons (Fliers, Noppen, Wiersinga, Visser, & Swaab, 1994) and glutamic acid decarboxylase (GAD) 65 and 67 (Gao & Moore, 1996a, 1996b). This supports the possible homology with the SDN in the rat, in which these peptide- and GABA-containing neurons have also been described (Bloch, Eckersell, & Mills, 1993; Gao & Moore, 1996a, 1996b). Moreover, some scattered *substance-P* neurons are present in the human SDN-POA (Chawla, Gutierrez, Scott Young, Mc-Mullen, & Rance, 1997), and moderate substance-P cell numbers were found in this area in rat (Simerly, Gorski, & Swanson, 1986).

Although it seems that all the data favor the homology between the human and the rat SDN-POA and are supported in this by the study of Koutcherov et al. (2002), it should be noted that others have claimed, on the basis of the presence of a sex difference, a possible homology between the rat SDN-POA (Gorski et al., 1978) and the human INAH-3 (Allen et al., 1989b; Byne et al., 2000; LeVay, 1991). However, this claim did not take into account a homology in neuropeptide and GABA content of those nuclei. One of the main arguments against homology between the human and rat SDN-POA has been the more lateral location of the human nucleus. In a very detailed developmental study, Koutcherov et al. (2002) have, however, given an excellent ontogenetic explanation for this seeming discrepancy. By means of immunocytochemistry, my colleagues and I recently found a more intense staining in men than in women for the androgen receptor (the receptor for testosterone) and for the estrogen receptors α and β in the SDN-POA, further supporting the presence of a sex difference in this nucleus (Fernández-Guasti, Kruijver, Fodor, & Swaab, 2000; Kruijver, Balesar, Espila, Unmehopa, & Swaab, 2002, 2003). Concerning the presence of estrogen receptors in the SDN-POA, it is relevant to note that an estrogen response element has been found within the human galanin gene (Howard, Peng, & Hyde, 1997).

Development and Sexual Differentiation of the SDN-POA

Sexual dimorphism does not seem to be present in the human SDN-POA at the time of birth. At that moment, total cell numbers are still similar in boys and girls and the SDN-POA contains no more than some 20% of the total cell number found between 2 to 4 years of age. From birth up to this age, cell numbers increase equally rapidly in both sexes (Figure 3.5). A sex difference in the SDN-POA does not occur until about the 4th year postnatally, when cell numbers start to decrease in girls, whereas in men the cell numbers in the SDN-POA remain stable until their rapid decrease at approximately 50 years of age, by which time the sex difference disappears. In women a second phase of marked cell loss sets in after the age of 70 (Figure 3.4; Hofman & Swaab, 1989; Swaab & Hofman, 1988), by which time the sex difference emerges again. The sharp decrease in cell numbers in the SDN-POA later in life may be related to the dramatic hormonal changes that accompany both male and female senescence (Hofman & Swaab, 1989) and to the decrease in male sexual activity around age 50 (Vermeulen, 1990). However, it is not clear whether the hormonal changes are directly related to these changes in various functions, either as cause or as effect of the observed cell loss in this nucleus.

The sex difference in the pattern of aging, and the fact that sexual differentiation in the human SDN-POA only occurs after the 4th year of age (Swaab & Hofman, 1988; Figure 3.5), might explain why Allen et al. (1989b), who had a sample of human adults biased for aged individuals, did not find a significant sex difference in the size of the SDN-POA, which they called INAH-1. In Allen et al.'s (1989b) study, 40% of the adult subjects came from the age group in which the SDN sex difference is minimal, compared with 29% in our study (Hofman & Swaab, 1989). Moreover, the age group of elderly subjects (over 70 years of age) was underrepresented in Allen et al.'s (1989b) study: 20%, compared with the 37.5% that would be a proportional distribution of all ages. In our study, 32% of the subjects belonged to this old age group. So it seems likely that Allen et al. (1989b) were unable to establish a sex difference in the INAH-1 (SDN-POA) because they used a biased sample. A further argument for this assumption is that, if we, in our material, had studied only subjects of the age distribution of Allen et al.'s (1989b) study, the sex difference in SDN-POA volume would have been reduced from a factor 2 (Hofman & Swaab, 1989) to only 1.4 times, and this difference would no longer have been statistically significant. Moreover, the sex difference in the SDN-POA emerges only between the ages of 4 and puberty (Swaab & Hofman, 1988);

therefore the brain of the 5-year-old boy and 4-year-old girl (she indeed had by far the largest volume of the entire series of female INAH-1) also produced a bias in the Allen et al. (1989b) study material. The age distribution, however, does not explain why LeVay (1991) and Byne et al. (2000) could not confirm the presence of a sex difference in the volume of INAH-1. Although the numbers of subjects they studied were much smaller than those in our study (Swaab & Hofman, 1988), technical differences such as section thickness may be a possible explanation for the controversy. Anyhow, the finding that nuclear androgen (Figure 3.7) and estrogen receptors α and β staining in the SDN-POA were more intense in men than in women (Fernández-Guasti et al., 2000; Kruijver et al., 2002, 2003) supports the presence of a functional sex difference in this nucleus.

A prominent theory is that sexual orientation develops as a result of an interaction between the developing brain and sex hormones (Dörner, 1988; Gladue, Green, & Helleman, 1984). According to Dörner's hypothesis, homosexual men would have a female differentiation of the hypothalamus. Although LeVay's (1991) data on the small female-sized INAH-3 in homosexual men are in agreement with this theory, this idea was not supported by our data on the SDN-POA in homosexual men. Neither the SDN-POA volume nor the cell number of homosexual men (Figure 3.6) who had died of AIDS differed from that of the male reference groups in the same age range or from that of heterosexuals who also have AIDS (Swaab & Hofman, 1988, 1990). The fact that no difference in SDN-POA cell number was observed between homosexual and heterosexual men, and the large SCN found in homosexual men (Swaab & Hofman, 1990), refutes the general formulation of Dörner's (1988) hypothesis that homosexual men would have "a female hypothalamus" and rather favors the idea that homosexual men have a hypothalamus that is different from the hypothalamus of both heterosexual men and heterosexual women. No data are available as yet about the hypothalamus in lesbian women.

Other Hypothalamic Sexual Dimorphisms

A number of other structural and functional hypothalamic sexual dimorphisms have been reported.

Structural Differences

In addition to the sex differences observed in the SCN, in the SDN-POA, and in the BST, Allen et al. (1989b) described two other cell groups (the interstitial nuclei of the anterior hypothalamus, INAH-2 and -3) that were larger in the male brain than in the female brain (see Figure 3.1). Because nothing is known about their neurotransmitter content, it is not clear at present which nuclei in the rat or rhesus monkey are homologous

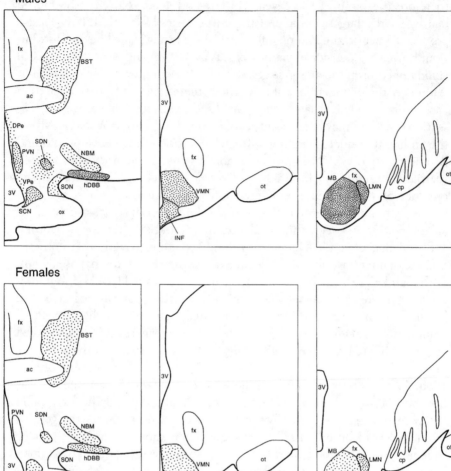

Figure 3.7. Schematic representation of the sex differences in the intensity of androgen receptor immunoreactivity in the human hypothalamus. The denser the dots, the more intensely stained the receptors for testosterone. Where there are no dots, no receptors were observed. NBM = nucleus basalis of Meynert; hDBB = horizontal limb of the diagonal band of Broca; SDN = sexually dimorphic nucleus of the preoptic area; SCN = suprachiasmatic nucleus; BST = bed nucleus of the stria terminalis; PVN = paraventricular nucleus; SON = supraoptic nucleus; DPe = periventricular nucleus dorsal zone; VPe = periventricular nucleus ventral zone; fx = fornix; ox = optic chiasma; 3V = third ventricle; ac = anterior commissure; VMN = ventromedial hypothalamic nucleus; INF = infundibular nucleus; ot = optic tract; MB = mamillary body (i.e., MMN [medial mamillary nucleus] + LMN [lateromamillary nucleus]); cp = cerebral peduncle. From "Sex Differences in the Distribution of Androgen Receptors in the Human Hypothalamus," by A. Fernández-Guasti, F. P. M. Kruijver, M. Fodor, and D. F. Swaab, 2000, *Journal of Comparative Neurology, 425,* p. 426. Reprinted with permission.

to the human INAH-2 and -3. There is a discrepancy in the literature concerning the sex difference in the size of INAH-2 as described by Allen et al. (1989b) that could not be confirmed by LeVay (1991) or by Byne et al. (2000). The fact that LeVay did not observe a smaller INAH-2 in women was proposed to be explained by an age-related sex difference in this nucleus. INAH-2 shows this sex difference only after the child-bearing age, with one exception: a 44-year-old woman who had a hysterectomy with ovarian removal 3 years prior to her death and who had a small INAH-2 (Allen et al., 1989b). The sex difference in INAH-2 thus seems to come to expression only after menopause, when circulating estrogens are absent. This would also explain why LeVay could not confirm the difference in INAH-2 in his group of young patients. The sex difference in INAH-2 was considered to be the first human example of a sex difference depending on circulating levels of sex hormones, that is, a difference based on a lack of activating effects of sex hormones in menopause rather than on organizing effects of sex hormones in development. However, Byne et al. (2000) could not confirm the relationship between INAH-2 and reproductive status as suggested by the data of Allen et al. (1989b).

Sex Hormone Receptor Distribution Differences

In most hypothalamic areas that contain androgen receptor staining, nuclear staining in particular is less intense in women than in men (Figure 3.7).The strongest sex difference was found in the lateral and the *medial mamillary nucleus* (MMN; Fernández-Guasti et al., 2000). The mamillary body complex is known to be involved in several aspects of sexual behavior, such as arousal of sexual interest and penile erection (Fernández-Guasti et al., 2000; MacLean & Ploog, 1962). In addition, a sex difference in androgen receptor staining was present in the horizontal diagonal band of Broca, SDN-POA, mPOA, the dorsal and ventral zone of the periventricular nucleus (PVN), SON, ventromedial hypothalamic nucleus, and the *infundibular nucleus*. However, no sex differences were observed in androgen receptor staining in the BST, the nucleus basalis of Meynert, and the island of Calleja (Fernández-Guasti et al., 2000). Nuclear androgen receptor activity in the mamillary complex of heterosexual men did not differ from that of homosexual men, but it was significantly stronger than in women. A femalelike pattern was found in 26- and 53-year-old castrated men and in intact old men. These data indicate that the amount of nuclear receptor staining in the mamillary complex is dependent on the circulating levels of androgens rather than on gender identity or sexual orientation. This idea is supported by the finding that a malelike pattern of androgen-receptor staining was found in a 36-year-old bisexual noncastrated male-to-female transsexual and a heterosexual virilized woman of 46 years of age (Kruijver, Fernández-

Guasti, Fodor, Kraan, & Swaab, 2001). Various sex differences were observed for estrogen receptor α (ERα) staining in the hypothalamus and adjacent areas of young human subjects. More intense nucleus ERα immunoreactivity was found in young men compared with young women, for example, in the SDN-POA, the SON, and the PVN. Women showed a stronger nuclear ERα-immunoreactivity in the SCN and MMN. No sex differences in nuclear ERα staining were found in, for example, the BSTc, the islands of Calleja (Cal), or in the infundibular nucleus (INF).

A more intense nuclear ERβ staining was found in men, for example, in neurons of the BSTc, the islands of Calleja, and the SDN-POA. Women showed more nuclear ERβ staining in the SCN, the SON, the PVN, the INF, and the MMN.

Observations in subjects with abnormal hormone levels showed, in most areas, ERβ-immunoreactivity distribution patterns according to the level of circulating estrogens, suggesting that the majority of the reported sex differences in ERβ-immunoreactivity are "activational" rather than "organizational" in nature (Kruijver et al., 2002, 2003).

Sex Differences in the Supraoptic and Paraventricular Nucleus

Not only does vasopressin act as an antidiuretic hormone, it is also involved in penile erection (Segarra et al., 1998; see below). Although my colleagues and I did not find a sex difference in vasopressin neuron number in the SON or PVN, a sex difference was reported in vasopressin plasma levels. Men have higher vasopressin levels than women (Asplund & Åberg, 1991; Van Londen et al., 1997). In addition, it was found that the posterior lobe of the pituitary is larger in boys than in girls (Takano, Utsunomiya, Ono, Ohfu, & Okazaki, 1999). This sex difference is explained by the higher activity we found in vasopressin neurons in the SON in young men as compared with young women as determined by the size of the Golgi apparatus. In the course of aging, probably triggered by the decrease in estrogen levels, the neuronal activity gradually increases in women, but it remains stable in men. The sex difference in neuronal activity in the SON thus disappears after the age of 50 (Ishunina, Salehi, Hofman, & Swaab, 1999). This is an example of a hypothalamic system that does not show a structural sex difference but rather a functional sex difference. It is also an example of a sex difference based on the activating (or in this case inhibiting) effect of sex hormones in adulthood.

The activation of neurosecretory vasopressin neurons in postmenopausal women was confirmed by in situ hybridization (Ishunina, Salehi, & Swaab, 2000) and by measuring the cell size as a parameter for neuronal activity. The vasopressin neurons in the SON and PVN appeared to be larger in young men than in young women. In elderly women (>50 years old), vasopressin cell size considerably exceeded that of young women. In addition, vasopressin cell size correlated positively with age in women but not in men. Sex differences in the size of the PVN vasopressin neurons were pronounced at the left side and absent at the right side, indicating the presence of functional lateralization of this nucleus. No difference was found in any morphometric parameter of oxytocin neurons in the PVN among the four groups studied. These data demonstrate sex differences in the size of the vasopressin neurons, and thus presumably in their function, that are age- and probably also side-dependent, and the absence of such changes in oxytocin neurons in the PVN (Ishunina & Swaab, 1999). The activation of vasopressin neurons in postmenopausal women is probably mediated by a decrease in the presence of estrogen receptor-β in these neurons and an increase in estrogen receptor α nuclear staining (Ishunina, Kruijver, Balesar, & Swaab, 2000).

The low-affinity neurotrophin receptor p75 (p75[NTR]) may be involved in the mechanism of activation of vasopressin neurons in postmenopausal women. This receptor was found to be expressed in the SON neurons of aged individuals, whereas p75[NTR] expression was shown to be suppressed by estrogens in a cell line. My colleagues and I investigated whether p75[NTR] immunoreactivity in SON neurons was age- and sex-dependent in postmortem brains of control patients ranging in age from 29 to 94 years old with an anti-p75[NTR] antibody and determined the area of p75[NTR] immunoreactivity per neuron profile using an image analysis system. To study whether the p75[NTR] might also participate in the activation of SON neurons, we related Golgi apparatus size to the amount of p75[NTR] in the same patients. P75[NTR] immunoreactivity indeed correlated significantly with age and with Golgi-apparatus size, but only in women, not in men. These observations suggest that p75[NTR] modulates the effects of estrogens on vasopressinergic neurons in the human SON (Ishunina, Salehi, & Swaab, 2000).

Oxytocin and vasopressin are thought to be involved in affiliation, including pair bonding, parental care, and territorial aggression, in monogamous animals (Insel, 1997; Young, Wang, & Insel, 1998); maternal behavior; and other aspects of reproductive behavior (Anderson-Hunt & Dennerstein, 1995; L. S. Carter, 1992; Insel, 1992; McKenna, 1998). Human sexually dimorphic reactions to pheromones (Savic, Berglund, Gulyas, & Roland, 2001) may affect such processes in the PVN. Lesions in the male rat PVN also indicate that this nucleus is involved in erection and that the magnocellular and parvicellular elements play different parts in this function (Liu et al., 1997). Electrical stimulation of the PVN in squirrel monkeys elicited penile erection (MacLean & Ploog, 1962). Electrical stimulation of the dorsal penile nerve or of the glans penis excited 60% of the oxytocin cells in the contralateral and ipsilateral SON of the rat (Honda et al., 1999). In another experiment, too, oxytocin cells of the PVN were found to be activated by sensory information from the penis. In the male rat, oxytocin is an extremely potent inducer of penile erection (Argiolas, Melis, & Gessa, 1986; McKenna, 1998). Because intracerebroventricular injection of an oxytocin antagonist reduced noncontact penile erections in rat dose-dependently, the involvement of central oxytocin in the expression of penile erections does not only seem to be a pharmacological effect but to have a physiological function as well (Melis, Spano, Succu, & Argiolas, 1999). The direct contractile effects of vasopressin on penile blood vessels, together with its amplifying effects on adrenergic-mediated constriction, support the idea that circulating vasopressin may also be involved in penile erection (Segarra et al., 1998). Tactile stimulation of the penis during male copulatory behavior further activates oxytocin cells, both in the PVN and in the SON, and this seems to induce both central and peripheral oxytocin release. Excitatory aminoacid transmission increases in the PVN during noncontact erections. This may contribute to the nitric-oxide production in the PVN and activates oxytocin neurons, thus mediating this sexual response (Melis, Succu, Spano, & Argiolas, 2000). Dopamine neurotransmission to the PVN is also supposed to be involved in penile erection (Chen, 2000). In addition, oxytocin may reduce maternal aggression in a period shortly after the birth, when lactating females show naturally high levels of this behavior (Giovenardi, Padoin, Cadore, & Lucion, 1998). In women, basal levels of oxytocin during lactation are associated with a desire to please, give, and interact socially (Uvnäs-Moberg, Alster, Petersson, Sohlström, & Björkstrand, 1998). Oxytocin released during labor and lactation may influence human maternal responsiveness and perhaps attachment (C. S. Carter, 1998). The increased oxytocin levels in cerebrospinal fluid during labor in human are presumed to be associated with the induction of maternal behavior (Takeda, Kuwabara, & Mizuno, 1985).

In a number of cases milk "let-down," indicating oxytocin release, has been reported during the sexual act in women who were in their lactating period (Campbell & Petersen, 1953). In men, oxytocin may also be involved in sexual arousal and ejaculation (Carmichael et al., 1987; Murphy, Seckl, Burton, Checkley, & Lightman, 1987). In women, too, oxytocin secretion seems to be related to smooth muscle contractions during orgasm (Carmichael et al., 1987). One minute after orgasm, oxytocin levels are increased in women (Blaicher et al., 1999). In agreement with this case report, oxytocin levels are known to rise during sexual arousal and peak during orgasm in both women and men. In multiorgasmic subjects, oxytocin peaks immediately prior to and during terminative orgasm, that is, the oxytocin peak coincides with sexual satiation. The intensity of orgasmic contractions, but not their duration, correlate positively with increases in oxytocin levels. Naloxon decreases the level of pleasure at orgasm and blocks the periorgasmic rise of oxytocin levels (Carmichael, Warburton, Dixen, & Davidson, 1994;

Murphy, Checkley, Seckl, & Lightman, 1990). Administration of vasopressin inhibits copulatory behavior in female rats, whereas a vasopressin antagonist facilitated the lordosis response (Meyerson, Höglund, Johansson, Blomqvist, & Ericson, 1988). Both in men and women, oxytocin induces contractions of smooth muscle cells and may thus facilitate transport of eggs and sperm (Carmichael et al., 1994). Animal experiments in rat confirm the possibility of such roles (Ackerman, Lange, & Clemens, 1998). In female animals, oxytocin was found to facilitate estrus, sexual arousal, receptivity, and other mating behaviors, including lordosis. An as yet unconfirmed case report has described a woman who began to take a contraceptive pill containing progesterone only. She experienced accentuated physiological and psychological sexual arousal after she had coincidentally used a prescribed synthetic oxytocin spray for let-down of breast milk (Anderson-Hunt, 1994; Anderson-Hunt & Dennerstein, 1995).

Animal experiments show a mechanism of interaction between sex hormones and oxytocin by initiating the production of receptors for this peptide (Anderson-Hunt & Dennerstein, 1995). Oxytocin is also produced by the male reproductive tract and modulates not only its contractility but also steroidogenesis. The finding that the oxytocin receptor is present in the interstitial tissue and in Sertoli cells in the testes supports the presence of such biological actions of oxytocin (Frayne & Nicholson, 1998). Indeed, oxytocin levels are significantly higher in women on oral contraceptives, and oxytocin levels recorded during the menstrual week are significantly lower than at other times (Uvnäs-Moberg, Sjögren, Westlin, Andersson, & Stock, 1989). Oxytocin plasma levels increase on estrogen administration (Kostoglou-Athanassiou, Athanassio, Treacher, Wheeler, & Forsling, 1998). The plasma concentration of vasopressin during the menstrual cycle is doubled on Days 16 to 18 as compared with Day 1 (Forsling, Åkerlund, & Strömberg, 1981), whereas others found a tendency of vasopressin to increase on Days 11 to 13, when peak concentrations of estradiol occur (Punnonen et al., 1983). Not only were the basal vasopressin levels higher in the follicular phase of the natural menstrual cycle, their nocturnal peaks were also higher (Kostoglou-Athanassiou, Athanassiou, Treacher, Wheeler, & Forsling, 1998). Increased oxytocin plasma levels were reported in a few women around the time of ovulation (Mitchell, Haynes, Anderson, & Turnbull, 1980).

TRANSSEXUALITY AND THE BED NUCLEUS OF THE STRIA TERMINALIS

Transsexuality is a rare condition. The annual incidence of transsexuality in Sweden has been estimated to be 0.17 per 100,000 inhabitants. The sex ratio (genetic male:female) varies from country to country between 1.4:1 and 3:1 (Garrels et al., 2000; Landén, Wålinder, & Lundström, 1996). There is only little information about the factors that may influence gender and cause transsexuality in humans (see Table 3.1). For gender identity disorder early in development, a strong (62%) hereditary component was found on the basis of twin studies (Coolidge, Thede, & Young, 2002). The disparate maternal aunt–uncle ratio in transsexual men has been hypothesized to be due to genomic imprinting, that is, a different effect depending on whether the gene is paternally or maternally inherited (Green & Keverne, 2000). There are only a few reports that have found chromosomal abnormalities in transsexuals. Six cases of male-to-female transsexuals with 47,XYY chromosome and one female-to-male transsexual with 47,XXX have been reported (Turan et al., 2000). Moreover, transsexualism has been reported in a man with Kleinfelter (XXY). In addition, pairs of monozygotic female twins have requested sex reassignment, and twin studies and familial cases of gender identity problems suggest a genetic basis for this disorder (Coolidge et al., 2002; Green, 2000; Sadeghi & Fakhrai, 2000). Although only a minority of the transsexuals has an underlying endocrine abnormality (Meyer et al., 1986), there are some indications of a possible disorder of the hypothalamo-pituitary gonadal axis in some transsexuals that may have a basis in development, such as a high frequency of polycystic ovaries, oligomenorrhea, and amenorrhea in female-to-male transsexualism (Futterweit, Weiss, & Fagerstrom, 1986; Sadeghi & Fakhrai, 2000).

Dessens et al. (1999) reported that three children born of a group of 243 women exposed to the anticonvulsants phenobarbital and diphantoin were found to be transsexuals, while, in addition, there were a few other subjects with gender dysphoria/cross-gender behavior. Gender problems thus occurred remarkably often in view of the rarity of this disorder. This exciting observation on the effect of compounds that are known to alter steroid levels in animal experiments has to be examined further. In this respect, it is of interest to note that phenobarbital has been widely used as prophylactic treatment in neonatal jaundice and greatly elevated the postnatal rise in testosterone (Forest, Lecog, Salle, & Bertrand, 1981). In 1996, Meyer-Bahlburg et al. reported a gender change from woman to man in four 46,XX individuals with classical congenital adrenal hyperplasia (CAH). Congenital adrenal hyperplasia, characterized by high androgen levels during prenatal development, indeed constitutes a risk factor for the development of gender identity problems (Slijper et al., 1998). Although it should be emphasized that the large majority of women with this disorder do not experience a marked gender identity conflict, the odds ratio that a genetic female with this disease would live, as an adult, in the male social role compared with genetic females in the general population was found to be 608:1 (Zucker et al., 1996). These observations support the view that intrauterine or perinatal exposure to abnormal levels of sex hormones may permanently affect gender identity.

The concept of sexual neutrality at birth after which infants differentiate as masculine or feminine as a result of social experiences was proposed by Money, Hampson, and Hampson (1955a, 1955b). Gender imprinting was presumed to start at the age of 1 year and to be already well established by 3 to 4 years of age (Money & Erhardt, 1972). Observations on children with male pseudohermaphroditism due to 5-α-reductase-2 deficiency were supposed to support the influence of life experience on psychosexual makeup (Al-Attia, 1996). A classic report that has strongly influenced the opinion that the environment plays a crucial role in gender development was the one described by Money of a boy whose penis was accidentally ablated at the age of 8 months during a phemosis repair by cautery and who was subsequently raised as a girl. Orchiectomy followed within a year to facilitate feminization. This individual was initially described as developing into a normally functioning woman. Later, however, it appeared that the individual had rejected the sex of raising and switched at puberty to living as a man again, and he requested male hormone shots, a mastectomy, and phalloplasty. At the age of 25 he married a woman and adopted her children. This famous John-Joan-John story, although it is just one case, illustrates that there is little, if any, support for the view that individuals are sexually neutral at birth and that normal psychosexual development is dependent on the environment of the genitals (Colapinto, 2001; Diamond & Sigmundson, 1997). This unfortunate story had a very sad ending. "John" committed suicide in May 2004.

In a second case of penile ablation in which the decision was made to reassign the patient as a girl and raise the baby as a girl, the remainder of the penis and testes were removed at a slightly earlier stage, at 7 months. Although her adult sexual orientation was bisexual and she was mainly attracted to women, her gender identity was female. The different outcome as compared with the former case is explained by the authors on the basis of the earlier decision to reassign the sex (Bradley, Oliver, Chernick, & Zucker, 1998). A third case, which did not get much attention, was a boy who lost his penis at the age of 6 months. A sex reassignment to the female gender was performed. However, in adolescence the patient refused hormone treatment and requested reassigment to a boy. Phalloplasty was performed by court order (Ochoa, 1998).

Male patients with cloacal exstrophy have a herniation of the urinary bladder and hindgut, and the anatomy leaves them aphalic in the majority of cases, although the testicles are normal from a histological point of view.

In a group of 8 male patients who were gender reassigned as females and orchidectomized in the neonatal period, at least in 3 instances gender identity has been questioned by the patients themselves (Zderic, Canning, Carr, Kodman-Jones, & Snyder, 2002), supporting the early programming of gender identity by biological factors and arguing against a dominating role of the social environment. The observation that in longitudinal series of 16 hormonally normal 46XY males assigned to female sex-of-rearing at birth, due to the absence of a penis, 8 have spontaneously declared themselves male and 15 fall very close to the male-typical spectrum of gender roles (Reiner, 2002), leads to the same conclusion. In a follow-up of this study, 8 of the 14 male subjects who were sex-reassigned as females declared themselves male. This study also indicates that the prenatal androgens are the major biological factor for the development of male gender identity, even in the absence of the neonatal and pubertal androgen surges (Reiner & Gearhart, 2004).

Reiner (1996) described a 46XY child with mixed gonadal dysgenesis, one immature testis, hypoplastic uterus, and clitoral hypertrophy, who was raised without stigmatization as a girl but who declared himself male at the age of 14. Following corrective surgery and testosterone substitution, he lived as a boy despite the social factors that were clearly in favor of maintaining the assigned sex. Apparently the deficient testis had been able to organize the brain during development even though the hormone levels were prenatally so inadequate that ambiguity of the genitalia was induced.

A child with true hermaphroditism, 45X (13%) 47XYY (87%) sex chromosome mosaic pattern in blood, uterus, fallopian tubes, phallus, testicular tissue, and epididymis, was assigned the male sex at birth. At 5 weeks the decision was made to reassign him to female. At 9 months an operation was performed to make the genitalia female, at 13 months the testicle was removed, and at the age of 5 another operation was done to make the genitalia female. She was raised as a girl but had masculine interests, and when she was around 8 years of age she declared that "God had made a mistake" and that she "should have been a boy." The male sex hormones to which she had been exposed in utero had apparently imprinted the male gender, although the authors also presumed postnatal psychosocial factors to have played a role (Zucker, Bradley, & Hughes, 1987).

We recently found a female-sized BSTc in male-to-female transsexuals. This supports the hypothesis that gender identity develops as a result of an interaction between the developing brain and sex hormones. The BST is situated at the junction of the hypothalamus, septum, and amygdala (Lesur, Gaspar, Alvarez, & Berger, 1989; Walter, Mai, Lanta, & Görcs, 1991; Figure 3.1) and plays an essential part in rodent sexual behavior (Liu et al., 1997). Estrogen and androgen receptors have been found in the human BST (Fernández-Guasti et al., 2000; Kruijver et al., 2002), and it is a major aromatization center in the developing rat brain (i.e., converting androgens into estrogens). The BST in rat receives projections, mainly from the amygdala, and provides a strong input in the preoptic-hypothalamic region. Reciprocal connections between the hypothalamus, BST, and amygdala are well documented in experimental animals (Liu et al., 1997; Zhou, Hofman, Gooren, & Swaab, 1995). There is a strong innervation of galanin fibers in the BST, and galanin receptors have also been shown in this structure (Mufson et al., 1998). The BST and centromedial amygdala have common cyto- and chemo-architectonic characteristics, and these regions are considered to be two components of one distinct neuronal complex. Neurons in the substantia innominata form cellular bridges between the BST and amygdala (Heimer, 2000; Lesur et al., 1989; Martin, Powers, Dellovade, & Price, 1991; Walter et al., 1991). In most mammals, including humans, the extended amygdala presents itself as a ring of neurons encircling the internal capsule and the basal ganglia (Heimer, Harlan, Alheid, Garcia, & De Olmos, 1997). The BST–amygdala continuum contains, for example, leutinizing hormone-releasing hormone (LHRH) neurons (Rance et al., 1994). However, tyrosine-hydroxylase mRNA has been found by others in these structures (Gouras, Rance, Scott Young, & Koliatsos, 1992).

Five principal sectors have been identified in the BST, including a darkly staining posteromedial component (dspm) of the BST (Allen et al., 1989a; Kruijver et al., 2000; Lesur et al., 1989; Martin et al., 1991). This part of the BST is situated in the dorsolateral zone of the fornix (Figure 3.1) and is sexually dimorphic (Figure 3.6). This sex difference does not seem to occur before adulthood. Its chemical composition and relationship to the other four principal BST sectors (see above) is unknown. Although the BST contains nuclear androgen receptors, no sex difference in the staining of this receptor was observed (Fernández-Guasti et al., 2000; Figure 3.7).

The volume of the BST-dspm is 2.5 times larger in men than in women (Allen et al., 1989a). My colleagues and I have found that the BSTc, which was defined by its dense *vasoactive intestinal polypeptide* (VIP) innervation (Figure 3.6) or by its somatostatin fibers and neurons, is sexually dimorphic. The BSTc is 40% smaller in women than in men and contains also some 40% fewer somatostatin neurons (Figure 3.6). No relationship was observed between BSTc volume or somatostatin cell number and sexual orientation: In the heterosexual reference group and a group of homosexual men, a similar BSTc volume and somatostatin cell number were observed. The size and somatostatin cell number of the BSTc were, moreover, not influenced by abnormal hormone levels in adulthood. However, a remarkably small BSTc (40% of the male reference volume and somatostatin neuron number) was observed in a group of 6 male-to-female transsexuals (Figures 3.6 and 3.9). These data suggest that the female size of this nucleus in male-to-female transsexuals was established during development and that the BSTc is part of a network that might be involved in gender, that is, the feeling of being either a man or a woman (Kruijver et al., 2000; Zhou, Hofman, Gooren, & Swaab, 1995).

HOMOSEXUALITY AND HYPOTHALAMIC STRUCTURES

Sexual orientation is influenced by quite a number of genetic as well as nongenetic factors (see Table 3.1). Genetic factors appear from studies in families, in twins, and through molecular genetics (Bailey & Bell, 1993; Bailey et al., 1999; Hamer, Magnuson, Hu, & Pattatucci, 1993; Hu et al., 1995; Kallman, 1952; Pillard & Bailey, 1998; Turner, 1995). Hamer and colleagues found linkage between DNA markers on the X-chromosome and male sexual orientation. Genetic linkage between the microsatellite markers on the X chromosome (i.e., Xq28) was detected for the families of gay men but not for the families of lesbians (Hamer et al., 1993, Hu et al., 1995). In a follow-up study, Rice, Anderson, Risch, and Ebers (1999) studied the sharing of alleles at position Xq28 in 52 Canadian gay male sibling pairs. Allele and haplotype sharing for these markers was not increased more than expected, which did not support the presence of an X-linked gene underlying male homosexuality. In a reaction to Rice et al.'s article, Hamer (1999) stated that (a) the family pedigree data from the Canadian study supported his hypothesis, (b) three other available Xq28 DNA studies found linkage, and (c) the heritability of sexual orientation is supported by substantial evidence independent of the X-chromosome data. In a meta-analysis of the four available studies, Hamer found a significant linkage. Rice et al. responded extensively and remained convinced that an X-linked gene could not exist in the population with any sizable frequency. This controversy will undoubtedly continue for a while longer. Aromatase cytochrome P450 (CYP19), which is necessary for the conversion of androgen to estrogens, was studied as a candidate gene for male homosexuality in homosexual brothers. However, the study revealed no indications that variation in this gene may be a major factor for the development of male homosexuality (DuPree, Mustanski, Bocklandt, Nievergelt, & Hamer, 2004).

Sex hormones during development also have an influence on sexual orientation, as appears from the increased proportion of bisexual and homosexual girls with CAH (Dittmann, Kappes, & Kappes, 1992; Meyer-Bahlberg et al., 1996; Money, Schwartz, & Lewis, 1984; Zucker et al., 1996). Then there is DES, a compound related to estrogens that increases the occurrence of bisexuality or homosexuality in girls whose mothers received DES during pregnancy (Ehrhardt et al., 1985; Meyer-Bahlburg et al., 1995) to prevent miscarriage (which it does not do). Whether environmental estrogens from plastics can influence sexual differentiation of the human brain and behavior is, at present, in debate but certainly not established. In addition, phytoestrogens, such as resveratrol, present in grapes and wine and an agonist for the estrogen receptor, should be considered in this respect (Gehm, McAndrews, Chien, & Jameson, 1997).

Maternal stress is thought to lead to increased occurrence of homosexuality in boys (Ellis, Ames, Peckham, & Burke, 1988) and girls (Bailey, Willerman, & Parks, 1991), and nicotine prenatally increases the occurrence of lesbianism (Ellis & Cole-Harding, 2001). As an interesting case history of this prenatal environmental factor, Weyl (1987) mentioned that Marcel Proust's mother was subjected to the overwhelming stress of the Paris commune during the 5th month of her pregnancy in 1871 and that Mary, Queen of Scots, the mother of the homosexual king of England, James I, toward the end of the 5th month of pregnancy had the terrifying experience that her secretary and special friend Riccio was killed. Although postnatal social factors are generally presumed to be involved in the development of sexual orientation (Byne & Pearson, 1993; Zucker et al., 1996), solid evidence in support of such an effect has not yet been reported. The observation that children raised by lesbian couples or by transsexuals generally have a heterosexual orientation (Golombok, Spencer, & Rutter, 1983; Green, 1978; Kirkpatrick, Smith, & Roy, 1981) does not support the possibility of the social environment in which the child is raised as an important factor for determining sexual orientation, nor is there scientific support for the idea that homosexuality has psychoanalytical or other psychological or social learning explanations, or that it would be a "lifestyle choice" (Ellis, 1996).

Various hypothalamic structures are structurally different in relation to sexual orientation, that is, the SCN, INAH-3, and the commissura anterior, suggesting that a difference in hypothalamic neuronal networks that occurs in development may be the basis of differences in sexual orientation. In addition to its possible involvement in reproduction (see below), the SCN might also play a role in sexual orientation. In fact, the first difference in the human brain in relation to sexual orientation was observed in the SCN. Morphometric analysis of the SCN of 10 homosexual men revealed that the volume of this nucleus was 1.7 times larger than that of a reference group of 18 presumed heterosexual men, and that it contained 2.1 times as many cells (Figure 3.8; Swaab & Hofman, 1990). In fact, the same high number of SCN vasopressin neurons as observed in 1- to 2-year-old children (Swaab, Hofman, & Honnebier, 1990) were also found in homosexual men. It seems as if the programmed postnatal cell death, which seems to begin in the SCN between 13 to 16 months after birth, does not occur to the same extent in homosexual men. The increased number of vasopressin-expressing neurons in the SCN of homosexual men appeared to be quite specific for this subgroup of neurons, because the number of

SUPRACHIASMATIC NUCLEUS   SEXUALLY DIMORPHIC NUCLEUS

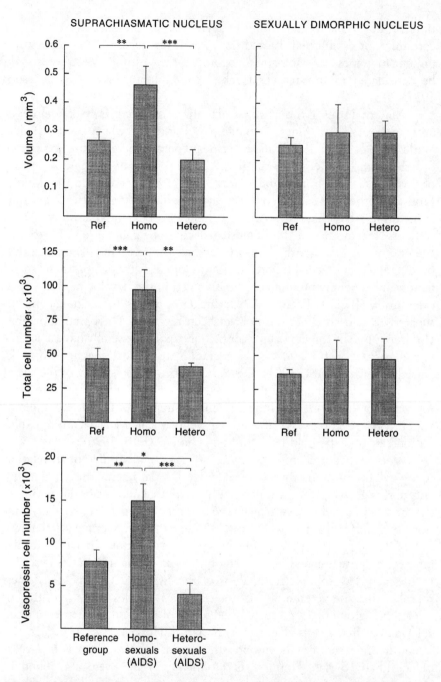

Figure 3.8. Top: Volume of the human suprachiasmatic nucleus (SCN) and sexually dimorphic nucleus of the preoptic area (SDN) as measured in three groups of adult subjects: a male reference group (Ref; *n* = 18); male homosexuals who died of AIDS (Homo; *n* = 10); heterosexuals who died of AIDS (Hetero; *n* = 6; 4 men and 2 women).

(caption continued on next page)

VIP-expressing neurons was not changed. However, in both the vasopressin and VIP neurons in the SCN, a reduced nuclear diameter was observed in homosexual men, suggesting metabolic alterations in the SCN in relation to sexual orientation (Zhou, Hofman, & Swaab, 1995a).

There are indeed a number of experimental data and observations on human material that indicate that the SCN is involved in aspects of sexual behavior and reproduction. Already in the early 1970s, postcoital ultrastructural changes indicating neuronal activation were reported in the SCN of the female rabbit (Clattenburg, Singh, & Montemurro, 1972). Important is also that the activity of SCN neurons increases suddenly around puberty (Anderson, 1981), indicating the addition of a reproductive function to the already mature circadian functions of the SCN. In addition, efferents of the SCN innervate the preoptic area that is involved in reproductive behaviors. Extensive lesioning of the SCN area results in failure of ovulation in the female rat (Brown-Grant & Raisman, 1977). The ovarian reproductive cycle is controlled by the SCN, probably through a direct monosynaptic innervation of LHRH neurons by VIP fibers (Van der Beek, Horvath, Wiegant, Van den Hurk, & Buijs, 1997; Van der Beek, Wiegant, Van der Donk, Van den Hurk, & Buijs, 1993). Several morphological sex differences have been reported that support putative reproductive functions. The SCN of male rats contains a larger amount of axo-spinal synapses, postsynaptic density material, and asymmetrical synapses, and their neurons contain more nucleoli than those of female rats (Güldner, 1982, 1983). The sex difference in synaptic number in the rat SCN depends on androgens in development (Le Blond, Morris, Karakiulakis, Powell, & Thomas, 1982). In gerbils the volume of the SCN is sexually dimorphic (Holman & Hutchison, 1991) and so is the organization of astroglia in the SCN (Collado, Beyer, Huchison, & Holman, 1995).

Figure 3.8. (caption continued from previous page). The values indicate medians and the standard deviation of the median. The differences in the volume of the SCN between homosexuals and the subjects from both other groups are statistically significant (Kruskal-Wallis multiple comparison test, $*p < .05$; $**p < .01$; $***p < .001$). Note that none of the parameters measured in the SDN (A,B) showed significant differences among the three groups (p always > 0.4). Middle: Total number of cells in the human SCN and SDN. The SCN in homosexual men contains 2.1 times as many cells as the SCN in the reference group of male subjects and 2.4 times as many cells as the SCN in heterosexual AIDS patients. Bottom: The number of vasopressin neurons in the human SCN (the SDN does not contain vasopressin-producing cells). The SCN in homosexual men contains, on average, 1.9 times as many vasopressin neurons as the SCN in heterosexual AIDS patients. Notice that the SCN of heterosexual individuals who died of AIDS contains fewer vasopressin cells than the SCN of the subjects from the reference group. From "An Enlarged Suprachiasmatic Nucleus in Homosexual Men," by D. F. Swaab and M. A. Hofman, 1990, *Brain Research, 537*, p. 144. Copyright 1990 by Elsevier. Reprinted with permission.

A sex difference was found in the shape of the vasopressin subdivision of the human SCN (Swaab, Fliers, & Partiman, 1985), and a sex difference was observed in the number of VIP-containing neurons in the human SCN. The number of VIP-expressing neurons in the SCN is larger in men of 10 to 40 years and larger in women of 41 to 65 years of age (Swaab, Zhou, Ehlhart, & Hofman, 1994; Zhou, Hofman, & Swaab, 1995b). These observations are also consistent with sexually dimorphic functions of the SCN that must, however, still be better defined. It is interesting to note, moreover, that the pineal hormone 5-methoxytryptophol shows significant sex differences: Plasma concentrations increase in boys and decrease in girls after the age of 8 (Molin-Carballo et al., 1996).

In seasonal breeders, VIP immunoreactivity in the SCN changes in relation to seasonal fluctuations in sexual activity (Lakhdar-Ghazal, Kalsbeek, & Pévet, 1992). The activation of c-fos in the SCN by sexual stimulation also points to a role of the SCN in reproduction (Pfaus, Kleopoulos, Mobbs, Gibbs, & Pfaff, 1993). Bakker, Van Ophemert, and Slob (1993) have found that male rats treated neonatally with the aromatase inhibitor 1,4,6androstatriene-3,17-dione (ATD) showed a clear sexual partner preference for females when tested in the late dark phase. When tested in the early dark phase, however, they showed a lesser preference for the female, or no preference at all. This is the first indication of the involvement of the clock (i.e., the SCN) in sexual orientation. The number of vasopressin-expressing neurons in the SCN of these ATD-treated bisexual animals was increased (Swaab et al., 1995), something that was also found in homosexual men (Swaab & Hofman, 1990). This observation supports the possibility that the increased number of vasopressin-expressing neurons in the SCN of adult homosexual men reflects a difference in the early stages of development in the brain.

Moreover, LeVay (1991) found that INAH-3 was twice as large in heterosexual men as in homosexual men. INAH-2 in the human hypothalamus is said to correspond to the anterocentral nucleus in the rhesus monkey and INAH-3 to the dorsocentral portion of the anterior hypothalamic nucleus in the rhesus monkey (Byne, 1998). Allen and Gorski (1992) found that the anterior commissure was larger in homosexual men than in heterosexual men.

CONCLUSIONS

Sex differences in the brain and in hormone levels are thought to be the structural and functional basis of pronounced sex differences of all aspects of sexual behavior. On the basis of animal experiments, sexual differentiation of the brain is thought to be imprinted or organized by androgens during

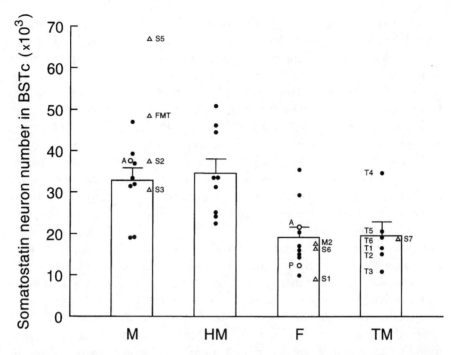

Figure 3.9. Central nucleus of the bed nucleus of the stria terminalis (BSTc) neuron numbers. Distribution of the BSTc neuron numbers among the different groups according to sex, sexual orientation, and gender identity: M (heterosexual male reference group), HM (homosexual male group), F (female group), TM (male-to-female transsexuals). The sex hormone disorder patients S1,2,3,5,6, and M2 indicate that changes in sex hormone levels in adulthood do not change the neuron numbers of the BSTc. The difference between the M and the TM group (<.04) becomes also statistically significant according to the sequential Bonferonni method if S2, S3, and S5 are included in the M group or if S7 is included in the TM group ($p \leq .01$). Note that the number of neurons of the female-to-male transsexual (FMT) is fully in the male range. A = AIDS patient. The BSTc number of neurons in the heterosexual man and woman with AIDS remained well within the corresponding reference group (see also Figure 3.1), so AIDS did not seem to affect the somatostatin neuron numbers in the BSTc. P = Postmenopausal woman; S1 (& 25 years of age) = Turner syndrome (45,X0; ovarian hypoplasia); M2 (& 73 years of age) = postmenopausal status. From "Male-to-Female Transsexuals Have Female Neuron Numbers in a Limbic Nucleus," by F. P. M. Kruijver, J. N. Zhou, C. W. Pool, M. A. Hofman, L. J. G. Gooren, and D. F. Swaab, 2000, *Journal of Clinical Endocrinology and Metabolism, 85,* p. 2036. Copyright 2000 by The Endocrine Society. Reprinted with permission.

development, following conversion by aromatase to estrogens, presumably during gestation or the perinatal period. From puberty onward, sex hormones alter the function of previously organized neuronal systems. This action is known as *activating* effects of sex hormones, but such effects may also inhibit neuronal function. However, genetic males with a complete androgen insensitivity syndrome, despite their normal testis differentiation and androgen biosynthesis, have a phenotype that is normal female, both external and in their behavior. They perceive themselves as highly feminine and do not have gender problems. This means that for the development of human male gender identity and male heterosexuality, direct androgen action on the brain seems to be of crucial importance. The aromatization theory may thus be of secondary importance for human sexual differentiation.

Apart from the interactions of hormones and the developing neurons, a number of factors seem to influence sexual differentiation of the brain and thus sexual orientation and gender identity. Chromosomal disorders, genetic factors, stress during pregnancy, and medicines taken by the pregnant mother may play a role. Postnatal social factors do not seem to be of primary importance for the development of sexual orientation or gender identity.

Focusing on the hypothalamus and adjacent areas, sex differences have been reported in volume and cell numbers, in neuronal metabolic activity of brain areas, and in their transmitter content that seem to be related to reproduction, gender identity, and sexual orientation. The BSTc, a brain area involved in reproduction, was found to be twice as large in men as in women. In male-to-female transsexuals, this sex difference was found to be reversed (Figure 3.9). The BSTc of the genetically male transsexuals was found to be of female size and neuron number, supported by the idea that this structure may be involved in the feeling to be a man or a woman. The BSTc in transsexuals thus seems to have followed a course of sexual differentiation during development that is opposite to that of the sex organs.

The interaction among hormones, hypothalamic structures, and the amygdala seem to be of the utmost importance for sexual behavior. Disorders of sexual behavior may not only go together with decreased hormone levels in elderly people but also be due to countless neurological, neuroendocrine, or psychiatric diseases and therefore demand careful diagnostic attention.

REFERENCES

Achermann, J. C., Meeks, J. J., & Jameson, J. L. (2001). Phenotypic spectrum of mutations in DAX-1 and SF-1. *Molecular and Cellular Endocrinology, 185,* 17–25.

Achermann, J. C., Weiss, J., Lee, E.-J., & Jameson, J. L. (2001). Inherited disorders of the gonadotropin hormones. *Molecular and Cellular Endocrinology, 179,* 89–96.

Ackerman, A. E., Lange, G. M., & Clemens, L. G. (1998). Effects of paraventricular lesions on sex behavior and seminal emission in male rats. *Physiology and Behavior, 63,* 49–53.

Al-Attia, H. M. (1996). Gender identity and role in a pedigree of Arabs with intersex due to 5 alpha reductase-2 deficiency. *Psychoneuroendocrinology, 21,* 651–657.

Albert, D. J., Walsh, M. L., & Jonk, R. H. (1993). Aggression in humans: What is its biological foundation? *Neuroscience and Biobehavioral Reviews, 17,* 405–425.

Allen, L. S., & Gorski, R. A. (1991). Sexual dimorphism of the anterior commissure and massa intermedia of the human brain. *Journal of Comparative Neurology, 312,* 97–104.

Allen, L. S., & Gorski, R. A. (1992). Sexual orientation and the size of the anterior commissure in the human brain. *Proceedings of the National Academy of Sciences USA, 89,* 7199–7202.

Allen, L. S., Hines, M., Shryne, J. E., & Gorski, R. A. (1989a). Sex difference in the bed nucleus of the stria terminalis of the human brain. *Journal of Comparative Neurology, 302,* 697–706.

Allen, L. S., Hines, M., Shryne, J. E., & Gorski, R. A. (1989b). Two sexually dimorphic cell groups in the human brain. *Journal of Neuroscience, 9,* 497–506.

Anderson, C. H. (1981). Nucleolus: Changes at puberty in neurons of the suprachiasmatic nucleus and the preoptic area. *Experimental Neurology, 74,* 780–786.

Anderson-Hunt, M. (1994). Increased female sexual response after oxytocin. *British Medical Journal, 309,* 929.

Anderson-Hunt, M., & Dennerstein, L. (1995). Oxytocin and female sexuality. *Gynecologic and Obstetric Investigation, 40,* 217–221.

Argiolas, A., Melis, M. R., & Gessa, G. L. (1986). Oxytocin: An extremely potent inducer of penile erection and yawning in male rats. *European Journal of Pharmacology, 130,* 265–272.

Arnow, B. A., Desmond, J. E., Banner, L. L., Glover, G. H., Solomon, A., Lake Polan, M., et al. (2002). Brain activation and sexual arousal in healthy, heterosexual males. *Brain, 125,* 1014–1023.

Asplund, R., & Åberg, H. (1991). Diurnal variation in the levels of antidiuretic hormone in the elderly. *Journal of Internal Medicine, 229,* 131–134.

Bailey, J. M., & Bell, A. P. (1993). Familiality of female and male homosexuality. *Behavior Genetics, 23,* 313–322.

Bailey, J. M., Pillard, R. C., Dawood, K., Miller, M. B., Farrer, L. A., Trivedi, S., & Murphy, R. L. (1999). A family history study of male sexual orientation using three independent samples. *Behavior Genetics, 29,* 79–86.

Bailey, J. M., Willerman, L., & Parks, C. (1991). A test of the maternal stress theory of human male homosexuality. *Archives of Sexual Behavior, 20,* 277–293.

Bakker, J., Van Ophemert, J., & Slob, A. K. (1993). Organization of partner preference and sexual behavior and its nocturnal rhythmicity in male rats. *Behavioral Neuroscience, 107,* 1–10.

Bancroft, J. (1999). Cardiovascular and endocrine changes during sexual arousal and orgasm. *Psychosomatic Medicine, 61,* 290–291.

Batch, J. A., Williams, D. M., Davies, H. R., Brown, B. D., Evans, B. A. J., Hughes, I. A., & Patterson, M. N. (1992). Role of the androgen receptor in male sexual differentiation. *Hormone Research, 38,* 226–229.

Bauer, H. G. (1954). Endocrine and other clinical manifestations of hypothalamic disease: A survey of 60 cases, with autopsies. *Journal of Clinical Endocrinology*, *14*, 13–31.

Bauer, H. G. (1959). Endocrine and metabolic conditions related to pathology in the hypothalamus: A review. *Journal of Nervous Mental Disease*, *128*, 323–338.

Blaicher, W., Gruber, D., Bieglmayer, C., Blaicher, A. M., Knogler, W., & Huber, J. C. (1999). The role of oxytocin in relation to female sexual arousal. *Gynecologic and Obstetric Investigation*, *47*, 125–126.

Bloch, G. J., Butler, P. C., & Kohlert, J. G. (1996). Galanin microinjected into the medial preoptic nucleus faciliates female- and male-typical sexual behaviors in the female rat. *Physiology and Behavior*, *59*, 1147–1154.

Bloch, G. J., Eckersell, C., & Mills, R. (1993). Distribution of galanin-immunoreactive cells within sexually dimorphic components of the medial preoptic area of the male and female rat. *Brain Research*, *620*, 259–268.

Bolton, N. J., Tapanainen, J., Koivisto, M., & Vihko, R. (1989). Circulating sex hormone-binding globulin and testosterone in newborns and infants. *Clinical Endocrinology*, *31*, 201–207.

Bonnefond, C., Palacios, J. M., Probst, A., & Mengod, G. (1990). Distribution of galanin mRNA containing cells and galanin receptor binding sites in human and rat hypothalamus. *European Journal of Neuroscience*, *2*, 629–637.

Braak, H., & Braak, E. (1987). The hypothalamus of the human adult: Chiasmatic region. *Anatomy and Embryology*, *176*, 315–330.

Braak, H., & Braak, E. (1992). Anatomy of the human hypothalamus (chiasmatic and tuberal region). In D. F. Swaab, M. A. Hofman, M. Mirmiran, R. Ravid, & F. W. Van Leeuwen (Eds.), *Progress in brain research: Vol. 93. The human hypothalamus in health and disease* (pp. 3–16). Amsterdam: Elsevier.

Bradley, S. J., Oliver, G. D., Chernick, A. B., & Zucker, K. J. (1998). Experiment of nurture: Ablatio penis at 2 months, sex reassignment at 7 months and a psychosexual follow-up in young adulthood. *Pediatrics*, *102*, 1–5.

Brockhaus, H. (1942). Beitrag zur normalen Anatomie des Hypothalamus und der Zona incerta beim Menschen. [Contribution to the normal anatomy of the human hypothalamus and zona inserta.] *Journal of Psychological Neurology*, *51*, 96–196.

Brown-Grant, K., & Raisman, G. (1977). Abnormalities in reproductive function associated with the destruction of the suprachiasmatic nuclei in female rats. *Proceedings of the Royal Society of London Series B*, *198*, 279–296.

Byne, W. (1998). The medial preoptic and anterior hypothalamic regions of the rhesus monkey: Cytoarchitectonic comparison with the human and evidence for sexual dimorphism. *Brain Research*, *793*, 346–350.

Byne, W., Lasco, M. S., Kemether, E., Shinwari, A., Edgar, M. A., Morgello, S., et al. (2000). The interstitial nuclei of the human anterior hypothalamus: An investigation of sexual variation in volume and cell size, number and density. *Brain Research*, *856*, 254–258.

Byne, W., & Pearson, B. (1993). Human sexual orientation: The biological theories reappraised. *Archives of General Psychiatry, 50,* 228–239.

Campbell, B., & Petersen, W. E. (1953). Milk "let-down" and the orgasm in the human female. *Human Biology, 25,* 165–168.

Carmichael, M. S., Humbert, R., Dixen, J., Palmisano, G., Greenleaf, W., & Davidson, J. M. (1987). Plasma oxytocin increases in the human sexual response. *Journal of Clinical Endocrinology and Metabolism, 64,* 27–31.

Carmichael, M. S., Warburton, V. L., Dixen, J., & Davidson, J. M. (1994). Relationships among cardiovascular, muscular, and oxytocin responses during human sexual activity. *Archives of Sexual Behavior, 23,* 59–79.

Carter, C. S. (1998). Neuroendocrine perspectives on social attachment and love. *Psychoneuroendocrinology, 23,* 779–818.

Carter, L. S. (1992). Oxytocin and sexual behavior. *Neuroscience and Biobehavioral Review, 16,* 131–144.

Chawla, M. K., Gutierrez, G. M., Scott Young, W., III, McMullen, N. T., & Rance, N. E. (1997). Localization of neurons expressing substance P and neurokinin B gene transcripts in the human hypothalamus and basal forebrain. *Journal of Comparative Neurology, 384,* 429–442.

Chen, K.-K. (2000). Paraventricular nucleus of hypothalamus: A brain locus in central neural regulation of penile erection in the rat. *International Journal of Andrology, 23*(Suppl. 2), 81.

Chung, W. C. J., De Vries, G. J., & Swaab, D. F. (2002). Sexual differentiation of the bed nucleus of the stria terminalis in humans may extend into adulthood. *Journal of Neuroscience, 22,* 1027–1033.

Clattenburg, R. E., Singh, R. P., & Montemurro, D. G. (1972). Postcoital ultrastructural changes in neurons of the suprachiasmatic nucleus of the rabbit. *Zeitschrift für Zellforschung, 125,* 448–459.

Cohen-Kettenis, P. T., & Gooren, L. J. G. (1998). Transsexualism: A review of etiology, diagnosis and treatment. *Journal of Psychosomatic Research, 46,* 315–333.

Colapinto, J. (2001). *As nature made him: The boy who was raised as a girl.* New York: Perennial.

Collado, P., Beyer, C., Hutchison, J. B., & Holman, S. D. (1995). Hypothalamic distribution of astrocytes is gender-related in Mongolian gerbils. *Neuroscience Letters, 184,* 86–89.

Coolidge, F. L., Thede, L. L., & Young, S. E. (2002). The heritability of gender identity disorder in a child and adolescent twin sample. *Behavior Genetics, 32,* 251–257.

Correa, R. V., Domenice, S., Bingham, N. C., Billerbeck, A. E. C., Rainey, W. E., Parker, K. L., & Mendonca, B. B. (2004). A microdeletion in the ligand binding domain of human steroidogenic factor 1 causes XY sex reversal without adrenal insufficiency. *Journal of Clinical Endocrinology and Metabolism, 89,* 1767–1772.

Cross, B. A., & Glover, T. D. (1958). The hypothalamus and seminal emission. *Journal of Endocrinology, 16,* 385–395.

De Jonge, F. H., Louwerse, A. L., Ooms, M. P., Evers, P., Endert, E., & Van de Poll, N. E. (1989). Lesions of the SDN-POA inhibit sexual behaviour of male Wistar rats. *Brain Research Bulletin, 23,* 483–492.

De Zegher, F., Devlieger, H., & Veldhuis, J. D. (1992). Pulsatile and sexually dimorphic secretion of luteinizing hormone in the human infant on the day of birth. *Pediatric Research, 32L,* 605–607.

Dessens, A. B., Cohen-Kettenis, P. T., Mellenbergh, G. J., Van de Poll, N. E., Koppe, J. G., & Boer, K. (1999). Prenatal exposure to anticonvulsants and psychosexual development. *Archives of Sexual Behavior, 28,* 31–44.

Dewing, P., Shi, T., Horvath, S., & Vilain, E. (2003). Sexually dimorphic gene expression in mouse brain precedes gonadal differentiation. *Molecular Brain Research, 118,* 82–90.

Diamond, M., & Sigmundson, K. (1997). Sex reassignment at birth: Long-term review and clinical implications. *Archives of Pediatric Adolescent Medicine, 151,* 298–304.

Dieckmann, G., Schneider-Jonietz, B., & Schneider, H. (1988). Psychiatric and neuropsychological findings after stereotactic hypothalamotomy, in cases of extreme sexual aggressivity. *Acta Neurochirurgica, 44*(Suppl.), 163–166.

Dittmann, R. W., Kappes, M. E., & Kappes, M. H. (1992). Sexual behavior in adolescent and adult females with congenital adrenal hyperplasia. *Psychoneuroendocrinology, 17,* 153–170.

Dörner, G. (1988). Neuroendocrine response to estrogen and brain differentiation in heterosexuals, homosexuals, and transsexuals. *Archives of Sexual Behavior, 17,* 57–75.

DuPree, M. G., Mustanski, B. S., Bocklandt, S., Nievergelt, C., & Hamer, D. H. (2004). A candidate gene study of CYPig (Aromatase) and male sexual orientation. *Behavior Genetics, 34,* 243–250.

Edwards, D. A., Walter, B., & Liang, P. (1996). Hypothalamic and olfactory control of sexual behavior and partner preference in male rats. *Physiology and Behavior, 60,* 1347–1354.

Ehrhardt, A. A., Meyer-Bahlburg, H. F. L., Rosen, L. R., Feldman, J. F., Veridiano, N. P., Zimmerman, I., & McEwen, B. S. (1985). Sexual orientation after prenatal exposure to exogenous estrogen. *Archives of Sexual Behavior, 14,* 57–75.

Ellis, L. (1996). Theories of homosexuality. In R. C. Savin-Williams & K. M. Cohen (Eds.), *The lives of lesbians, gays, and bisexuals: Children to adults* (pp. 11–70). Ft. Worth, TX: Harcourt Brace College.

Ellis, L., Ames, M. A., Peckham, W., & Burke, D. (1988). Sexual orientation of human offspring may be altered by severe maternal stress during pregnancy. *Journal of Sex Research, 25,* 152–157.

Ellis, L., & Cole-Harding, S. (2001). The effects of prenatal stress, and of prenatal alcohol and nicotine exposure, on human sexual orientation. *Physiology and Behavior, 74,* 213–226.

Exton, M. S., Bindert, A., Krüger T., Scheller, F., Hartmann, U., & Schedlowski, M. (1999). Cardiovascular and endocrine alterations after masturbation-induced orgasm in women. *Psychosomatic Medicine, 61*, 280–289.

Fernández-Guasti, A., Kruijver, F. P. M., Fodor, M., & Swaab, D. F. (2000). Sex differences in the distribution of androgen receptors in the human hypothalamus. *Journal of Comparative Neurology, 425*, 422–435.

Fliers, E., Noppen, N. W. A. M., Wiersinga, W. M., Visser, T. J., & Swaab, D. F. (1994). Distribution of thyrotropin-releasing hormone (TRH)-containing cells and fibers in the human hypothalamus. *Journal of Comparative Neurology, 350*, 311–323.

Forest, M. G. (1989). Physiological changes in circulating androgens. *Pediatric and Adolescent Endocrinology, 19*, 104–129.

Forest, M. G., Lecoq, A., Salle, B., & Bertrand, J. (1981). Does neonatal phenobarbital treatment affect testicular and adrenal functions and steroid binding in plasma in infancy? *Journal of Clinical Endocrinology and Metabolism, 52*, 103–110.

Forsling, M. L., Åkerlund, M., & Strömberg, P. (1981). Variations in plasma concentrations of vasopressin during the menstrual cycle. *Journal of Endocrinolgy, 89*, 263–266.

Frayne, J., & Nicholson, H. D. (1998) Localization of oxytocin receptors in the human and macaque monkey male reproductive tracts: Evidence for a physiological role of oxytocin in the male. *Molecular Human Reproduction, 4*, 527–532.

Frohman, E. M., Frohman, T. C., & Moreault, A. M. (2002). Acquired sexual paraphilia in patients with multiple sclerosis. *Archives of Neurology, 59*, 1006–1010.

Futterweit, W., Weiss, R. A., & Fagerstrom, R. M. (1986). Endocrine evaluation of forty female-to-male transsexuals: Increased frequency of polycystic ovarian disease in female transsexualism. *Archives of Sexual Behavior, 15*, 69–77.

Gai, W. P., Geffen, L. B., & Blessing, W. W. (1990). Galanin immunoreactive neurons in the human hypothalamus: Colocalization with vasopressin-containing neurons. *Journal of Comparative Neurology, 298*, 265–280.

Gao, B., & Moore, R. Y. (1996a). Glutamic acid decarboxylase message isoforms in human suprachiasmatic nucleus. *Journal of Biological Rhythms, 11*, 172–179.

Gao, B., & Moore, R. Y. (1996b). The sexually dimorphic nucleus of the hypothalamus contains GABA neurons in rat and man. *Brain Research, 742*, 163–171.

Garrels, L., Kockott, G., Michael, N., Preuss, W., Renter, K., Schmidt, G., et al. (2000). Sex ratio of transsexuals in Germany: The development over three decades. *Acta Psychiatrica Scandinavica, 102*, 445–448.

Gaus, S. E., Strecker, R. E., Tate, B. A., Parker, R. A., & Saper, C. B. (2002). Ventrolateral preoptic nucleus contains sleep-active, galaninergic neurons in multiple mammalian species. *Neuroscience, 115*, 285–294.

Gehm, B. D., McAndrews, J. M., Chien, P. Y., & Jameson, J. L. (1997). Resveratrol, a polyphenolic compound found in grapes and wine, is an agonist for the estrogen receptor. *Proceedings of the National Academy of Sciences USA, 94,* 14138–14143.

Giovenardi, M., Padoin, M. J., Cadore, L. P., & Lucion, A. B. (1998). Hypothalamic paraventricular nucleus modulates maternal aggression in rats: Effects of ibotenic acid lesion and oxytocin antisense. *Physiology and Behavior, 63,* 351–359.

Giuliano, F., Rampin, O., Brown, K., Courtois, F., Benoit, G., & Jardin, A. (1996). Stimulation of the medial preoptic area of the hypothalamus in the rat elicits increases in intracavernous pressure. *Neuroscience Letters, 209,* 1–4.

Gladue, B. A., Green, R., & Helleman, R. E. (1984, September 28). Neuroendocrine response to estrogen and sexual orientation. *Science, 225,* 1496–1499.

Golombok, S., Spencer, A., & Rutter, M. (1983). Children in lesbian and single-parent households: Psychosexual and psychiatric appraisal. *Journal of Child Psychology and Psychiatry, 4,* 551–572.

Gooren, L. J. G., Fliers, E., & Courtney, K. (1990). Biological determinants of sexual orientation. *Annual Review of Sex Research, 1,* 175–196.

Gorman, D. G., & Cummings, J. L. (1992). Hypersexuality following septal injury. *Archives of Neurology, 49,* 308–310.

Gorski, R. A., Gordon, J. H., Shryne, J. E., & Southam, A. M. (1978). Evidence for a morphological sex difference within the medial preoptic area of the rat brain. *Brain Research, 148,* 333–346.

Gottfries, C. G., Roos, B. E., & Winblad, B. (1974). Determination of 5-hydroxy-tryptamine, 5-hydroxyindoleacetic acid and homovanillic acid in brain tissue from an autopsy material. *Acta Psychiatrica Scandinavica, 50,* 496–507.

Gouras, G. K., Rance, N. E., Scott Young, W., III, & Koliatsos, V. E. (1992). Tyrosine-hydroxylase-containing neurons in the primate basal forebrain magnocellular complex. *Brain Research, 584,* 287–293.

Green, R. (1978). Sexual identity of 37 children raised by homosexual or transsexual parents. *American Journal of Psychiatry, 135,* 692–697.

Green, R. (2000). Family cooccurrence of "gender dysphoria": Ten sibling or parent–child pairs. *Archives of Sexual Behavior, 29,* 499–507.

Green, R., & Keverne, E. B. (2000). The disparate maternal aunt–uncle ratio in male-transsexuals: An explanation invoking genomic imprinting. *Journal of Theoretical Biology, 202,* 55–63.

Güldner, F.-H. (1982). Sexual dimorphisms of axo-spine synapses and postsynaptic density material in the suprachiasmatic nucleus of the rat. *Neuroscience Letters, 28,* 145–150.

Güldner, F.-H. (1983). Numbers of neurons and astroglial cells in the suprachiasmatic nucleus of male and female rats. *Experimental Brain Research, 50,* 373–376.

Hamer, D. H. (1999, August 6). Genetics and male sexual orientation. *Science, 285,* 803a.

Hamer, D. H., Magnuson, V. L., Hu, N., & Pattatucci, A. M. L. (1993, July 16). A linkage between DNA markers on the X chromosome and male sexual orientation. *Science, 261*, 321–327.

Heath, R. G. (Ed.). (1964). *The role of pleasure in behavior*. New York: Hoeber Medical Division, Harper & Row.

Heimann, H. (1979). Psychiatrische, psychologische, soziologische und ethische implikationen psychochirurgischer Maßnahmen unter besonderer Berücksichtigung der Hypothalamotomie bei Sexualdeviationen [Psychiatric, psychological, sociological, and ethical implications of psychosurgery in particular of hypothalectomy in sexual deviation]. *Nervenartz, 50*, 682–688.

Heimer, L. (2000). Basal forebrain in the context of schizophrenia. *Brain Research Review, 31*, 205–235.

Heimer, L., Harlan, R. E., Alheid, G. F., Garcia, M. M., & De Olmos, J. (1997). Substantia innominata: A notion which impedes clinical–anatomical correlations in neuropsychiatric disorders. *Neuroscience, 76*, 957–1006.

Hengstschläger, M., Van Trotsenburg, M., Repa, C., Marton, E., Hüber, J. C., & Bernaschek, G. (2003). Sex chromosome aberrations and transsexualism. *Fertility and Sterility, 79*, 639–640.

Hiort, O. (2000). Neonatal endocrinology of abnormal male sexual differentiation: Molecular aspects. *Hormone Research, 53*, 38–41.

Hofman, M. A., & Swaab, D. F. (1989). The sexually dimorphic nucleus of the preoptic area in the human brain: A comparative morphometric study. *Journal of Anatomy, 164*, 55–72.

Holman, S. D., & Hutchison, J. B. (1991). Differential effects of neonatal castration on the development of sexually dimorphic brain areas in the gerbil. *Developmental Brain Research, 61*, 147–150.

Honda, K., Yanagimoto, M., Negoro, H., Narita, K., Murata, T., & Higuchi, T. (1999). Excitation of oxytocin cells in the hypothalamic supraoptic nucleus by electrical stimulation of the dorsal penile nerve and tactile stimulation of the penis in the rat. *Brain Research Bulletin, 48*, 309–313.

Howard, G., Peng, L., & Hyde, J. F. (1997). An estrogen receptor binding site within the human galanin gene. *Endocrinology, 138*, 4649–4656.

Hu, S., Pattatucci, A. M. L., Patterson, C., Li, L., Fulker, D. W., Cherny, S. S., et al. (1995). Linkage between sexual orientation and chromosome Xq28 in males but not in females. *Nature Genetics, 11*, 248–256.

Imperato-McGinley, J., Miller, M., Wilson, J. D., Peterson, R. E., Shackleton, C., & Gajdusek, D. C. (1991). A cluster of male pseudohermaphrodites with 5α-reductase deficiency in Papua New Guinea. *Clinical Endocrinology, 34*, 293–298.

Imperato-McGinley, J., Peterson, R. E., Gautier, T., & Sturla, E. (1979). Androgens and the evolution of male-gender identity among male pseudohermaphrodites with 5α-reductase deficiency. *New England Journal of Medicine, 300*, 1233–1237.

Insel, T. R. (1992). Oxytocin—a neuropeptide for affiliation: Evidence from behavioral, receptor autoradiographic, and comparative studies. *Psychoneuroendocrinology, 17,* 3–35.

Insel, T. R. (1997). A neurobiological basis of social attachment. *American Journal of Psychiatry, 154,* 726–735.

Ishunina, T. A., Kruijver, F. P., Balesar, R., & Swaab, D. F. (2000). Differential expression of estrogen receptor alpha and beta immunoreactivity in the human supraoptic nucleus in relation to sex and aging. *Journal of Clinical Endocrinology and Metabolism, 85,* 3283–3291.

Ishunina, T. A., Salehi, A., Hofman, M. A., & Swaab, D. F. (1999). Activity of vasopressinergic neurons of the human supraoptic nucleus is age and sex dependent. *Journal of Neuroendocrinology, 11,* 251–258.

Ishunina, T. A., Salehi, A., & Swaab, D. F. (2000). Sex- and age-related p75 neurotrophin receptor expression in the human supraoptic nucleus. *Neuroendocrinology, 374,* 243–251.

Ishunina, T. A., & Swaab, D. F. (1999). Vasopressin and oxytocin neurons of the human supraoptic and paraventricular nucleus; size changes in relation to age and sex. *Journal of Clinical Endocrinology and Metabolism, 84,* 4637–4644.

Ishunina, T. A., & Swaab, D. F. (2002). Neurohyposeal peptides in aging and Alzheimer's disease. *Aging Research Reviews, 1,* 537–558.

Kallmann, F. J. (1952). Comparative twin study on the genetic aspects of male homosexuality. *Journal of Nervous Mental Diseases, 115,* 283–298.

Karama, S., Roch Lecours, A., Leroux, J.-M., Bourgouin, P., Beaudoin, G., Joubert, S., & Beauregard, M. (2002). Areas of brain activation in males and females during viewing of erotic film excerpts. *Human Brain Mapping 16,* 1–13.

Kartha, K. N. B., & Ramakrishna, T. (1996). The role of sexually dimorphic medial preoptic area of the hypothalamus in the sexual behaviour of male and female rats. *Physiological Research, 45,* 459–466.

Kindon, H. A., Baum, M. J., & Paredes, R. J. (1996). Medial preoptic/anterior hypothalamic lesions induce a female-typical profile of sexual partner preference in male ferrets. *Hormones and Behavior, 30,* 514–527.

Kirkpatrick, M., Smith, C., & Roy, R. (1981). Lesbian mothers and their children: A comparative survey. *American Journal of Orthopsychiatry, 51,* 545–551.

Kostoglou-Athanassiou, I., Athanassiou, K., Treacher, D. F., Wheeler, M. J., & Forsling, M. L. (1998). Neurohypophysial hormone and melatonin secretion over the natural and suppressed menstrual cycle in premenopausal women. *Clinical Endocrinology, 49,* 209–216.

Koutcherov, Y., Mai, J. K., Ashwell, K. W. S., & Paxinos, G. (2002). Organization of human hypothalamus in fetal development. *Journal of Comparative Neurology, 446,* 301–324.

Kruijver, F. P. M., Balesar, R., Espila, A. M., Unmehopa, U. A., & Swaab, D. F. (2002). Estrogen receptor-α distribution in the human hypothalamus in relation to sex and endocrine status. *Journal of Comparative Neurology, 454,* 115–139.

Kruijver, F. P. M., Balesar, R., Espila, A. M., Unmehopa, U. A., & Swaab, D. F. (2003). Estrogen-receptor β distribution in the human hypothalamus: Similarities and differences with ERα distribution. *Journal of Comparative Neurology, 466,* 251–277.

Kruijver, F. P. M., Fernández-Guasti, A., Fodor, M., Kraan, E., & Swaab, D. F. (2001). Sex differences in androgen receptors of the human mamillary bodies are related to endocrine status rather than to sexual orientation or transsexuality. *Journal of Clinical Endocrinology and Metabolism, 86,* 818–827.

Kruijver, F. P. M., Zhou, J. N., Pool, C. W., Hofman, M. A., Gooren, L. J. G., & Swaab, D. F. (2000). Male-to-female transsexuals have female neuron numbers in a limbic nucleus. *Journal of Clinical Endocrinology and Metabolism, 85,* 2034–2041.

Lakhdar-Ghazal, N., Kalsbeek, A., & Pévet, P. (1992). Sexual dimorphism and seasonal variations in vasoactive intestinal peptide immunoreactivity in the suprachiasmatic nucleus of jerboa (*Jaculus orientalis*). *Neuroscience Letters, 144,* 29–33.

Landén, M., Wålinder, J., & Lundström, B. (1996). Incidence and sex ratio of transsexualism in Sweden. *Acta Psychiatrica Scandinavica, 93,* 261–263.

Le Blond, C. B., Morris, S., Karakiulakis, G., Powell, R., & Thomas, P. J. (1982). Development of sexual dimorphism in the suprachiasmatic nucleus of the rat. *Journal of Endocrinology, 95,* 137–145.

Lesur, A., Gaspar, P., Alvarez, C., & Berger, B. (1989). Chemoanatomic compartments in the human bed nucleus of the stria terminalis. *Neuroscience, 32,* 181–194.

LeVay, S. (1991, August 30). A difference in hypothalamic structure between heterosexual and homosexual men. *Science, 253,* 1034–1037.

Lilly, R., Cummings, J. L., Benson, D. F., & Frankel, M. (1983). The human Klüver-Bucy syndrome. *Neurology, 33,* 1141–1145.

Liu, Y.-C., Salamone, J. D., & Sachs, B. D. (1997). Impaired sexual response after lesions of the paraventricular nucleus of the hypothalamus in male rats. *Behavioral Neuroscience, 111,* 1361–1367.

Macke, J. P., Hu, N., Hu, S., Bailey, M., King, V. L., Brown, T., et al. (1993). Sequence variation in the androgen receptor gene is not a common determinant of male sexual orientation. *American Journal of Human Genetics, 53,* 844–852.

MacLean, P. D., & Ploog, D. W. (1962). Cerebral representation of penile erection. *Journal of Neurophysiology, 25,* 29–55.

MacNeil, D. J., Howard, A. D., Guan, X., Fong, T. M., Nargund, R. P., Bednarek, M. A., et al. (2002). The role of melanocortins in body weight regulation: Opportunities for the treatment of obesity. *European Journal of Pharmacology, 440,* 141–157.

Marlowe, W. B., Mancall, E. L., & Thomas, J. J. (1975). Complete Klüver-Bucy syndrome in man. *Cortex, 11*, 53–59.

Martin, L. J., Powers, R. E., Dellovade, T. L., & Price, D. L. (1991). The bed nucleus–amygdala continuum in human and monkey. *Journal of Comparative Neurology, 309*, 445–485.

Mayer, A., Lahr, G., Swaab, D. F., Pilgrim, C., & Reisert, I. (1998). The Y-chromo-somal genes SRY and ZFY are transcribed in adult human brain. *Neurogenetics, 1*, 281–288.

McKenna, K. E. (1998). Central control of penile erection. *International Journal of Impotence Research, 10*, S25–S34.

Melis, M. R., Spano, M. S., Succu, S., & Argiolas, A. (1999). The oxytocin antago-nist d(CH$_2$)$_5$Tyr(Me)^2-Orn^8-vasotocin reduces non-contact penile erections in male rats. *Neuroscience Letters, 265L*, 171–174.

Melis, M. R., Succu, S., Spano, M. S., & Argiolas, A. (2000). Effect of excitatory amino acid, dopamine, and oxytocin receptor antagonists on noncontact penile erections and paraventricular nitric oxide production in male rats. *Behavioral Neuroscience 114*, 849–857.

Merari, A., & Ginton, A. (1975). Characteristics of exaggerated sexual behavior induced by electrical stimulation of the medial preoptic area in male rats. *Brain Research, 86*, 97–108.

Meyer, W. J., Webb, A., Stuart, C. A., Finkelstein, J. W., Lawrence, B., & Walker, P. A. (1986). Physical and hormonal evaluation of transsexual patients: A longitudinal study. *Archives of Sexual Behavior, 15*, 121–138.

Meyer-Bahlburg, H. F. L., Ehrhardt, A. A., Rosen, L. R., Gruen, R. S., Veridiano, N. P., Van, F. H., & Neuwalder, H. F. (1995). Prenatal estrogens and the development of homosexual orientation. *Developmental Psychology, 31*, 12–21.

Meyer-Bahlburg, H. F. L., Gruen, R. S., New, M. I., Bell, J. J., Morishima, A., Shimshi, M., et al. (1996). Gender change from female to male in classical congenital adrenal hyperplasia. *Hormones and Behavior, 30*, 319–332.

Meyers, R. (1961). Evidence of a locus of the neural mechanisms for libido and penile potency in the septo-fornico-hypothalamic region of the human brain. *Transactions of the American Neurological Association, 86*, 81–85.

Meyerson, B. J., Höglund, U., Johansson, C., Blomqvist, A., & Ericson, H. (1988). Neonatal vasopressin antagonist treatment facilitates adult copulatory behavior in female rats and increases hypothalamic vasopressin content. *Brain Research, 473*, 344–351.

Miller, B. L., Cummings, J. L., McIntyre, H., Ebers, G., & Grode, M. (1986). Hypersexuality or altered sexual preference following brain injury. *Journal of Neurology, Neurosurgery, and Psychiatry, 49*, 867–873.

Mitchell, M. D., Haynes, P. J., Anderson, A. B. M., & Turnbull, A. C. (1980). Oxytocin in human ovulation. *Lancet, 2*(8196), 704.

Mizusawa, H., Hedlund, P., & Andersson, K.-E. (2002). α-Melanocyte stimulating hormone and oxytocin induced penile erections, and intracavernous pressure increases in the rat. *Journal of Urology, 167*, 757–760.

Molin-Carballo, A., Muñoz-Hoyos, A., Martin-García, J. A., Uberos-Fernandéz, J., Rodriguez-Cabezas, T., & Acuña-Castroviejo, D. (1996). 5-Methoxytryptophol and melatonin in children: Differences due to age and sex. *Journal of Pineal Research, 21,* 73–79.

Molinoff, P. B., Shadiack, A. M., Earle, D., Diamond, L. E., & Quon, C. Y. (2003). PT-141: A melanocortin agonist for the treatment of sexual dysfunction. *Annals of the New York Academy of Sciences, 994,* 96–102.

Money, J., & Erhardt, A. A. (1972). *Man and woman, boy and girl: The differentiation and dimorphism of gender identity from conception to maturity.* Baltimore: Johns Hopkins University Press.

Money, J., Hampson, J. G., & Hampson, J. L. (1955a). An examination of some basic sexual concepts: The evidence of human hermaphroditism. *Bulletin of Johns Hopkins Hospital, 97,* 301–319.

Money, J., Hampson, J. G., & Hampson, J. L. (1955b). Hermaphroditism: Recommendations concerning assignment of sex, change of sex and psychological management. *Bulletin of Johns Hopkins Hospital, 97,* 284–300.

Money, J., Schwartz, M., & Lewis, V. G. (1984). Adult erotosexual status and fetal hormonal masculinization: 46,XX congenital virilizing adrenal hyperplasia and 46,XY androgen-insensitivity syndrome compared. *Psychoneuroendocrinology, 9,* 405–414.

Morishima, A., Grumbach, M. M., Simpson, E. R., Fisher, C., & Qin, K. (1995). Aromatase deficiency in male and female siblings caused by a novel mutation and the physiological role of estrogens. *Journal of Clinical Endocrinology and Metabolism, 80,* 3689–3698.

Mufson, E. J., Kahl, U., Bowser, R., Mash, D. C., Kordower, J. H., & Deecher, D. C. (1998). Galanin expression within the basal forebrain in Alzheimer's disease. *Annals of the New York Academy of Sciences, 863,* 291–304.

Müller, D., Roeder, F., & Orthner, H. (1973). Further results of stereotaxis in the human hypothalamus in sexual deviations. First use of this operation in addiction to drugs. *Neurochirurgia, 16,* 113–126.

Murphy, M. R., Checkley, S. A., Seckl, J. R., & Lightman, S. L. (1990). Naloxone inhibits oxytocin release at orgasm in man. *Journal of Clinical Endocrinology and Metabolism, 71,* 1056–1058.

Murphy, M. R., Seckl, J. R., Burton, S., Checkley, S. A., & Lightman, S. L. (1987). Changes in oxytocin and vasopressin secretion during sexual activity in men. *Journal of Clinical Endocrinology and Metabolism, 65,* 738–741.

Ochoa, B. (1998). Trauma of the external genitalia in children: Amputation of the penis and emasculation. *Journal of Urology, 160,* 1116–1119.

Ozisik, G., Achermann, J. C., & Jameson, J. L. (2002). The role of SF1 in adrenal and reproductive function: Insight from naturally occurring mutations in humans. *Molecular Genetics and Metabolism, 76,* 85–91.

Paredes, R. G., & Baum, M. J. (1995). Altered sexual partner preference in male ferrets given excitotoxic lesions of the preoptic area/anterior hypothalamus. *Journal of Neuroscience, 15,* 6619–6630.

Paredes, R. G., Tzschentke, T., & Nakach, N. (1998). Lesions of the medial preoptic area/anterior hypothalamus (MPOA/AH) modify partner preference in male rats. *Brain Research,* *813,* 1–8.

Pfaus, J. G., Kleopoulos, S. P., Mobbs, C. V., Gibbs, R. B., & Pfaff, D. W. (1993). Sexual stimulation activates c-fos within estrogen-concentrating regions of the female rat forebrain. *Brain Research, 624,* 253–267.

Pilgrim, C., & Reisert, I. (1992). Differences between male and female brains: Developmental mechanisms and implications. *Hormone and Metabolic Research, 24L,* 353–359.

Pillard, R. C., & Bailey, M. (1998). Human sexual orientation has a heritable component. *Human Biology, 70,* 347–365.

Poeck, K., & Pilleri, G. (1965). Release of hypersexual behaviour due to lesion in the limbic system. *Acta Neurologica Scandinavica, 41,* 233–244.

Punnonen, R., Viinamäki, O., & Multamäki, S. (1983). Plasma vasopressin during normal menstrual cycle. *Hormone Research, 17,* 90–92.

Quigley, A. (2002). The postnatal gonadotropin and sex steroid surge: Insights from the androgen insensitivity syndrome. *Journal of Clinical Endocrinology and Metabolism, 87,* 24–28.

Rance, N. E., Young, W. S., III, & McMullen, N. T. (1994). Topography of neurons expressing luteinizing hormone-releasing hormone gene transcripts in the human hypothalamus and basal forebrain. *Journal of Comparative Neurology, 339,* 573–586.

Reiner, W. G. (1996). Case study: sex reassignment in a teenage girl. *Journal of the American Academy of Child Adolescent Psychiatry, 35,* 799–803.

Reiner, W. G. (2002). Gender identity and sex assignments: A reappraisal for the 21st century. *Advanced Experiments in Medical Biology, 511,* 175–189.

Reiner, W. G., & Gearhart, J. P. (2004). Discordant sexual identity in some genetic males with cloacal exstrophy assigned to female sex at birth. *New England Journal of Medicine, 350,* 333–341.

Resko, J. A., Perkins, A., Roselli, C. E., Fitzgerald, J. A., Choate, J. V. A., & Stormshak, F. (1996). Endocrine correlates of partner preference behavior in rams. *Biology of Reproduction, 55,* 120–126.

Rice, G., Anderson, C., Risch, N., & Ebers, G. (1999, April 23). Male homosexuality: Absence of linkage to microsatellite markers at Xq28. *Science, 284,* 665–667.

Rieber, I., & Sigusch, V. (1979). Psychosurgery on sex offenders and sexual 'deviants' in West Germany. *Archives of Sexual Behavior, 8,* 523–527.

Roselli, C. E., Abdelgadir, S. E., & Resko, J. A. (1997). Regulation of aromatase gene expression in the adult rat brain. *Brain Research Bulletin, 44,* 351–357.

Roselli, C. E., Larkin, K., Resko, J. A., Stellflug, J. N., & Stormshak, F. (2004). The volume of a sexually dimorphic nucleus in the ovine medial preoptic area/anterior hypothalamus varies with sexual partner preference. *Endocrinology, 145,* 478–483.

Roselli, C. E., & Resko, J. A. (2001). Cytochrome P450 aromatase (CYP19) in the non-human primate brain: Distribution, regulation, and functional significance. *Journal of Steroid Biochemistry and Molecular Biology, 79*, 247–253.

Sadeghi, M., & Fakhrai, A. (2000). Transsexualism in female monozygotic twins: A case report. *Australian and New Zealand Journal of Psychiatry, 34*, 862–864.

Sasano, H., Takahashi, K., Satoh, F., Nagura, H., & Harada, N. (1998). Aromatase in the human central nervous system. *Clinical Endocrinology, 48*, 325–329.

Sato, T., Matsumoto, T., Kawano, H., Watanabe, T., Uematsu, Y., Sekine, K., et al. (2004). Brain masculinization requires androgen receptor function. *Proceedings of the National Academy of Sciences, 101*, 1673–1678.

Savic, I., Berglund, H., Gulyas, B., & Roland, P. (2001). Smelling of odorous sex hormone-like compounds causes sex-differentiated hypothalamic activations in humans. *Neuron, 30*, 661–668.

Schorsch, E., & Schmidt, G. (1979). Hypothalamotomie bei sexuellen Abweichungen [Hypothalectomy in case of sexual deviations]. *Nervenarzt, 50*, 689–699.

Schultz, C., Braak, H., & Braak, E. (1996). A sex difference in neurodegeneration of the human hypothalamus. *Neuroscience Letters, 212*, 103–106.

Segarra, G., Medina, P., Domenech, C., Vila, J. M., Martínez-León, J. B., Aldasoro, M., & Lluch, S. (1998). Role of vasopressin on adrenergic neurotransmission in human penile blood vessels. *Journal of Pharmacology and Experimental Therapeutics, 286*, 1315–1320.

Simerly, R. B., Gorski, R. A., & Swanson, L. W. (1986). Neurotransmitter specificity of cells and fibers in the medial preoptic nucleus: An immunohistochemical study in the rat. *Journal of Comparative Neurology, 246*, 343–362.

Skuse, D. H. (1999). Genomic imprinting of the X-chromosome: A novel mechanism for the evolution of sexual dimorphism. *Journal of Laboratory and Clinical Medicine, 133*, 23–32.

Slijper, F. M. E., Stenvert, L. S., Drop, M. D., Molenaar, J. C., & De Muinck Keizer-Schram, S. M. P. F. (1998). Long-term psychological evaluation of intersex children. *Archives of Sexual Behavior, 27*, 125

Smith, E. P., Boyd, J., Frank, G. R., Takahashi, H., Cohen, R. M., Specker, B., et al. (1994). Estrogen resistance caused by a mutation in the estrogen-receptor gene in a man. *New England Journal of Medicine, 331*, 1056–1061.

Swaab, D. F. (2002). Gender issues in brain structures and functions and their relevance for psychopathology. In H. D'haenen, J. A. Den Boer, & P. Willner (Eds.), *Biological psychiatry* (pp. 189–209). New York: Wiley.

Swaab, D. F., & Fliers, E. (1985, May 31). A sexually dimorphic nucleus in the human brain. *Science, 228*, 1112–1115.

Swaab, D. F., Fliers, E., & Partiman, T. S. (1985). The suprachiasmatic nucleus of the human brain in relation to sex, age and senile dementia. *Brain Research, 342*, 37–44.

Swaab, D. F., Gooren, L. J. G., & Hofman, M. A. (1992). The human hypothalamus in relation to gender and sexual orientation. In D. F. Swaab, M. A. Hofman, M. Mirmiran, R. Ravid, & F. W. Van Leeuwen (Eds.), *Progress in brain research: Vol. 93. The human hypothalamus in health and disease* (pp. 205–215). Amsterdam: Elsevier.

Swaab, D. F., & Hofman, M. A. (1984). Sexual differentiation of the human brain: A historical perspective. In De Vries et al. (Eds.), *Progress in brain research: Vol. 61. Sex differences in the brain: Relation between structure and function* (pp. 361–374). Amsterdam: Elsevier.

Swaab, D. F., & Hofman, M. A. (1988). Sexual differentiation of the human hypothalamus: Ontogeny of the sexually dimorphic nucleus of the preoptic area. *Developmental Brain Research, 44*, 314–318.

Swaab, D. F., & Hofman, M. A. (1990). An enlarged suprachiasmatic nucleus in homosexual men. *Brain Research, 537*, 141–148.

Swaab, D. F., & Hofman, M. A. (1995). Sexual differentiation of the human hypothalamus in relation to gender and sexual orientation. *Trends in Neurosciences, 18*, 264–270.

Swaab, D. F., Hofman, M. A., & Honnebier, M. B. O. M. (1990). Development of vasopressin neurons in the human suprachiasmatic nucleus in relation to birth. *Developmental Brain Research, 52*, 289–293.

Swaab, D. F., Kamphorst, W., Raadsheer, F. C., Purba, J. S., Ravid, R., & Tilders, F. J. H. (1995). Increased hypothalamo-pituitary-adrenal axis activity is not pivotal in the pathogenesis of Alzheimer's disease. In K. Iqbal, J. A. Mortimer, B. Winblad, & H. M. Wisniewski (Eds.), *Research advances in Alzheimer's disease and related disorders* (pp. 461–466). New York: Wiley.

Swaab, D. F., Zhou, J. N., Ehlhart, T., & Hofman, M. A. (1994). Development of vasoactive intestinal polypeptide Z(VIP) neurons in the human suprachiasmatic nucleus (SCN) in relation to birth and sex. *Developmental Brain Research, 79*, 249–259.

Takahashi, T., Shoji, Y., Shoji, Y., Haraguchi, N., Takahashi, I., & Takada, G. (1997). Active hypothalamic-pituitary-gonadal axis in an infant with X-linked adrenal hypoplasia congenita. *Journal of Pediatrics, 130*, 485–488.

Takano, K., Utsunomiya, H., Ono, H., Ohfu, M., & Okazaki, M. (1999). Normal development of the pituitary gland: assessment with three-dimensional MR volumetry. *American Journal of Neuroradiology, 20*, 312–315.

Takeda, S., Kuwabara, Y., & Mizuno, M. (1985). Effects of pregnancy and labor on oxytocin levels in human plasma and cerebrospional fluid. *Endocrinologia Japonica, 32*, 875–880.

Tatemoto, K., Rökaeus, Å., Jörnvall, H., McDonald, T. J., & Mutt, V. (1983). Galanin—A novel biologically active peptide from porcine intestine. *Federation of European Biochemical Societies, 164*, 124–128.

Terzian, H., & Dalle Ore, G. (1955). Syndrome of Klüver and Bucy: Reproduced in man by bilateral removal of the temporal lobes. *Neurology, 5*, 373–380.

Tucker Halpern, C., Udry, J. R., & Suchindran, C. (1998). Monthly measures of salivary testosterone predict sexual activity in adolescent males. *Archives of Sexual Behavior, 27,* 445–465.

Turan, M., Eşel, E., Dündar, M., Candemir, Z., Baştürk, M., Sofuoğlu, S., & Özkul, Y. (2000). Female-to-male transsexual with 47,XXX karyotype. *Biological Psychiatry, 48,* 1116–1117.

Turkenburg, J. L., Swaab, D. F., Endert, E., Louwerse, A. L., & Van de Poll, N. E. (1988). Effects of lesions of the sexually dimorphic nucleus on sexual behaviour of testosterone-treated female Wistar rats. *Brain Research Bulletin, 21,* 215–224.

Turner, W. J. (1995). Homosexuality, Type 1: An Xq28 phenomenon. *Archives of Sexual Behavior, 24,* 109–134.

Udry, J. R., Morris, N. M., & Kovenock, J. (1995). Androgen effects on women's gendered behaviour. *Journal of Biosocial Science, 27,* 359–368.

Uvnäs-Moberg, K., Alster, P., Petersson, M., Sohlström, A., & Björkstrand, E. (1998). Postnatal oxytocin injections cause sustained weight gain and increased nociceptive thresholds in male and female rats. *Pediatric Research, 43,* 344–348.

Uvnäs-Moberg, K., Sjögren, C., Westlin, L., Andersson, P. O., & Stock, S. (1989). Plasma levels of gastrin, somatostatin, VIP, insulin and oxytocin during the menstrual cycle in women (with and without oral contraceptives). *Acta Obstetrica et Gynecologica Scandinavica, 68,* 165–169.

Valdueza, J. M., Cristante L., Dammann, O., Bentele, K., Vortmeyer, A., Saeger, W., et al. (1994). Hypothalamic hamartomas: With special reference to gelastic epilepsy and surgery. *Neurosurgery, 34,* 949–958.

Van der Beek, E. M., Horvath, T. L., Wiegant, V. M., Van den Hurk, R., & Buijs, R. M. (1997). Evidence for a direct neuronal pathway from the suprachiasmatic nucleus to the gonadotropin-releasing hormone system: Combined tracing and light and electron microscopic immunocytochemical studies. *Journal of Comparative Neurology, 384,* 569–579.

Van der Beek, E. M., Wiegant, V. M., Van der Donk, H. A., Van den Hurk, R., & Buijs, R. M. (1993). Lesions of the suprachiasmatic nucleus indicate the presence of a direct VIP containing projection to gonadotropin-releasing hormone neurons in the female rat. *Journal of Neuroendocrinology, 5,* 137–144.

Van der Ploeg, L. H. T., Martin, W. J., Howard, A. D., Howard, A. D., Nargund, R. P., Austin, C. P., et al. (2002). A role for the melanocortin 4 receptor in sexual function. *Proceedings of the National Academy of Sciences, 99,* 11381–11386.

Van Eerdenburg, F. J. C. M., & Swaab, D. F. (1991). Increasing neuron numbers in the vasopressin and oxytocin containing nucleus of the adult female pig hypothalamus. *Neuroscience Letters, 132,* 85–88.

Van Londen, L., Goekoop, J. G., Van Kemper, G. M. J., Frankhuijsen-Sierevogel, A. C., Wiegant, V. M., Van der Velde, E. A., & De Wied, D. (1997). Plasma levels of arginine vasopressin elevated in patients with major depression. *Neuropsychopharmacology, 17,* 284–292.

Vermeulen, A. (1990). Androgens and male senescence. In E. Nieschlag & H. M. Behre (Eds.), *Testosterone: Action, deficiency, substitution* (pp. 629–645). Berlin, Germany: Springer Verlag.

Walter, A., Mai, J. K., Lanta, L., & Görcs, T. (1991). Differential distribution of immunohistochemical markers in the bed nucleus of the stria terminalis in the human brain. *Journal of Chemical Neuroanatomy, 4*, 281–298.

Ward, O. B. (1992). Fetal drug exposure and sexual differentiation of males. In A. A. Gerall, H. Moltz, & I. L. Ward (Eds.), *Handbook of behavioral neurobiology* (Vol. 11, pp. 181–219). New York: Plenum Press.

Watabe, T., & Endo, A. (1994). Sexual orientation of male mouse offspring prenatally exposed to ethanol. *Neurotoxicology and Teratology, 16*, 25–29.

Weyl, N. (1987). Hormonal influences on sexual inversion: A dual inheritance model of Proust's homosexuality. *Journal of Social and Biological Structures, 10*, 385–390.

Wilson, J. D. (1999). The role of androgens in male gender role behavior. *Endocrine Review, 20*, 726–737.

Wilson, J. D., Griffin, J. E., & Russell, D. W. (1993). Steroid 5α-reductase 2 deficiency. *Endocrinology Review, 14*, 577–593.

Wisniewski, A. B., Migeon, C. J., Meyer-Bahlburg, H. F. L., Gearhart, J. P., Berkovitz, G. D., Brown, T. R., & Money, J. (2000). Complete androgen insensitivity syndrome: Long-term medical, surgical, and psychosexual outcome. *Journal of Clinical Endocrinology and Metabolism, 85*, 2664–2669.

Yahr, P., Finn, P. D., Hoffman, N. W., & Sayag, N. (1994). Sexually dimorphic cell groups in the medial preoptic area that are essential for male sex behavior and the neural pathways needed for their effects. *Psychoneuroendocrinology, 19*, 463–470.

Young, L. J., Wang, Z., & Insel, T. R. (1998). Neuroendocrine bases of monogamy. *Trends in Neurosciences, 21*, 71–75.

Zderic, S. A., Canning, D. A., Carr, M. C., Kodman-Jones, C., & Snyder, HMcC (2002). The CHOP experience with cloacal exstrophy and gender reassignment. *Advanced Experiments in Medical Biology, 511*, 135–144.

Zhou, J. N., Hofman, M. A., Gooren, L. J. G., & Swaab, D. F. (1995). A sex difference in the human brain and its relation to transsexuality. *Nature, 378*, 68–70.

Zhou, J. N., Hofman, M. A., & Swaab, D. F. (1995a). No changes in the number of vasoactive intestinal polypeptide (VIP)-expressing neurons in the suprachiasmatic nucleus of homosexual men; comparison with vasopressin-expressing neurons. *Brain Research, 672*, 285–288.

Zhou, J. N., Hofman, M. A., & Swaab, D. F. (1995b). VIP neurons in the human SCN in relation to sex, age, and Alzheimer's disease. *Neurobiology of Aging, 16*, 571–576.

Zucker, K. J. (2002). Intersexuality and gender identity differentiation. *Journal of Pediatric Adolescent Gynecology, 15*, 3–13.

Zucker, K. J., Bradley, S. J., & Hughes, H. E. (1987). Gender dysphoria in a child with true hermaphroditism. *Canadian Journal of Psychiatry, 32*, 602–609.

Zucker, K. J., Bradley, S. J., Oliver, G., Blake, J., Fleming, S., & Hood, J. (1996). Psychosexual development of women with congenital adrenal hyperplasia. *Hormones and Behavior, 30*, 300–318.

GLOSSARY

Central nucleus of the bed nucleus of the stria terminalis (BSTc): a structure located between the ventral tip of the lateral ventricle and the anterior commissure. It is involved in many aspects of sexual behavior and is twice as large in males as in females. This sex difference is reversed in transsexuals. The sex difference in the BSTc thus seems to parallel gender identity. Many of the neurons of the BSTc contain the neuropeptide somatostatin as a messenger, and the BSTc receives a dense network of vasoactive intestinal polypeptide-containing fibers.

Galanin: a neuropeptide that is present in the SDN-POA neurons. This peptide induces both female and male types of sexual behavior and influences eating behavior and sleep. The production of this peptide is regulated by estrogens.

Gamma-amino butyric acid (GABA): the main inhibiting neurotransmitter in the brain, also present in the SDN-POA.

Infundibular nucleus: the same as the arcuate nucleus in rat. It is the area from where the menstrual cycle is regulated and a crucial structure for metabolism and growth.

Interstitial nucleus of the anterior hypothalamus (INAH 1–3): numbered 1–4 (see Figure 3.1). Small hypothalamic cell groups. INAH-1 is the same as the SDN-POA. The other INAH cell groups are chemically and functionally not defined.

Medial mamillary nucleus (MMN): the main component of the mamillary bodies, crucial for memory. These bodies receive information from the hippocampal complex by the fornix and transfer this information to the thalamus.

Morphometric analysis: a procedure to estimate—from thin sections under the microscope—total cell numbers, and so on, in a three-dimensional structure. The microscope is usually linked to a computer.

Paraventricular nucleus (PVN): the autonomic center of the hypothalamus, containing a large number of different neurons with different peptide content, different fields of termination, and different functions. The oxytocin and vasopressin neurons of the PVN are involved in affiliation, including pair bonding, parental care, territorial aggression, and aspects of sexual behavior, including penile erections.

Sexually dimorphic nucleus of the preoptic area (SDN-POA): a structure in the anterior hypothalamus that in young adults is about twice as large in males as in females. The medial preoptic area (mPOA) is involved in penile erection, mounting, intromission, ejaculation, and sexual orientation. In the literature this area is also called *INAH-1*, or intermediate nucleus.

Substance-P: a neuropeptide involved, for example, in pain perception. It is also present in small amounts in the SDN-POA.

Suprachiasmatic nucleus (SCN): the biological clock, situated on top of the crossing of the optic nerves. It regulates all day/night rhythms and appears to be twice as large in homosexual males as in heterosexual males. A major neuropeptide in the SCN is vasopressin.

Supraoptic nucleus (SON): situated on top of the optic nerve. It is responsible for the major part of circulating vasopressin. Vasopressin not only acts as antidiuretic hormone on the kidneys but is also involved in aspects of reproduction.

Thyrotropin-releasing hormone (TRH): a neuropeptide that is produced in the PVN and released into the portal capillaries in the median eminence and regulates, via the pituitary gland, the function of the thyroid. It also has central functions, such as metabolism and temperature regulation. It influences mood and is also present in the SDN-POA.

Vasoactive intestinal polypeptide (VIP): a peptide produced by some neurons as a chemical messenger. A dense network of VIP fibers innervates the BSTc.

4

THE CENTRAL CONTROL AND PHARMACOLOGICAL MODULATION OF SEXUAL FUNCTION

KEVIN E. McKENNA

In this review, I describe the central nervous system pathways that control sexual function. The peripheral control of sexual organs, especially the penis, has been extensively reviewed and is not reiterated here (Andersson & Wagner, 1995; De Groat & Booth, 1993; De Groat & Steers, 1990). I also review recent work on the pharmacological manipulation of various sexual responses. Most of the work reviewed has been performed in experimental animals, especially in rats. Therefore, considerable caution is needed in making conclusions of the effects of pharmacological agents on sexual function in humans. Furthermore, a large proportion of research has been devoted to studying penile erection in males. Functions such as ejaculation and female genital arousal have received considerably less attention. Some general conclusions can be drawn concerning the neural control of sexual function. Sexual reflexes are largely generated by neural pathways within the spinal cord. These can be activated by genital sensory stimulation. They are also under descending modulation from supraspinal centers. Both inhibitory and excitatory pathways have been identified. The excitatory pathways are capable of activating sexual responses even in the absence of genital stimulation, such as in the case of sexual arousal induced by mental imagery.

SPINAL INNERVATION OF THE SEXUAL ORGANS

In this section I discuss the various mechanisms of the nervous system that control sexual organ functioning through the spinal cord. First I discuss motor, sensory, and interneuron mechanisms, then I focus on spinal reflexes and the role of different parts of the brain.

Motor

Thorough reviews of the innervation of the pelvic organs have been published (Bell, 1972; Jänig & McLachlan, 1987). Pelvic innervation has been most extensively studied in the rat. In the rat, the autonomic innervation of the pelvic organs is provided primarily by the major pelvic ganglion (Langworthy, 1965; Purinton, Fletcher, & Bradley, 1976). This is a triangular structure located on the lateral surface of the cervix in females and the lateral lobe of the prostate in males. Damage to this ganglion, during prostate surgery in men and hysterectomy in women, can lead to sexual dysfunction (Hasson, 1993; Walsh & Donker, 2002). It is a mixed ganglion, containing both sympathetic and parasympathetic postganglionic neurons. Postganglionic fibers from the ganglion innervate the pelvic organs, including the bladder, urethra, accessory sex glands, vagina, uterus, clitoris, and penis. The cavernous nerve is the largest nerve issuing from the pelvic ganglion and provides the vasodilatory innervation to the penis and clitoris (Langworthy, 1965). After nearly a century of research, the neurotransmitter mediating the vasodilation of the penis and clitoris elicited by cavernous nerve activation has been identified as nitric oxide (Andersson & Wagner, 1995; Burnett, Lowenstein, Bredt, Chang, & Snyder, 1992; Park et al., 1997).

Application of tracers to the rat pelvic nerve has been used to label parasympathetic preganglionic neurons in the sacral parasympathetic nucleus in the lumbosacral segments of the spinal cord (Hancock & Peveto, 1979b; Nadelhaft & Booth, 1984). These neurons were located in a compact column in the lateral intermediate gray matter. This is on the border between the gray matter (nerve cells) and white matter (nerve fibers) of the spinal cord. No differences were noted in the number or distribution of labeled cells in males or females. Sympathetic preganglionic neurons in the thoracolumbar segments of the spinal cord were labeled following application of tracer to the hypogastric nerve (Hancock & Peveto, 1979a; Nadelhaft & McKenna, 1987). The preganglionic neurons were found bilaterally in the intermediolateral cell column and in the medial central gray, dorsal to the central canal.

The pudendal nerve innervates the striated perineal muscles, the external anal and urethral sphincters, and the ischiocavernosus and bulbospongiosus muscles (Breedlove & Arnold, 1981; Jordan, Breedlove, & Arnold, 1982; McKenna & Nadelhaft, 1986; Schrøder, 1980). The pudendal motoneurons are located in the lumbar spinal cord. The number of motoneurons is sexually dimorphic, with considerably more motoneurons in the male compared with the female. This directly reflects the dimorphism of the ischiocavernosus and bulbospongiosus muscles, which are prominent in the male and very small in the female rat.

Sensory

Anatomical studies have shown that sensory fibers from the pudendal, pelvic, and hypogastric nerves innervate pelvic organs. The pudendal nerve provides sensory innervation for the perineum, penis or clitoris, and urethra. In the female, the size and sensitivity of the pudendal perineal innervation are increased by estrogen (Adler, Davis, & Komisaruk, 1977; Komisaruk, Adler, & Hutchinson, 1972; Kow & Pfaff, 1973). The pudendal sensory innervation plays an essential role in sexual reflexes.

Pelvic nerve sensory fibers innervate the internal sexual organs. Individual fibers innervate only a single pelvic organ. Most vaginal and uterine afferent fibers appear to be unmyelinated (small fibers without an insulating sheath). In the female, the pelvic nerve innervation has been shown to be crucial for the induction of pregnancy or pseudopregnancy due to mating or cervical stimulation (Carlson & De Feo, 1965; Kollar, 1953). The role of pelvic nerve sensory innervation of the male sex organs is unclear. There are relatively few sensory fibers in the hypogastric nerve of the rat (Nadelhaft & McKenna, 1987). However, these neurons have been shown to be important for the pain sensation from the uterus (Berkley, Robbins, & Sato, 1987). The functions of hypogastric sensory neurons in the male are unknown.

The afferents from the pelvis terminate primarily in the medial portions of the dorsal horn and in the medial central gray matter of the lumbosacral spinal cord (McKenna & Nadelhaft, 1986; Nadelhaft & Booth, 1984; Nadelhaft & McKenna, 1987). The pudendal afferents have an almost exclusively medial distribution. Visceral pelvic afferents terminate both medially and in the lateral edge of the gray matter, in the vicinity of the sacral parasympathetic nucleus.

It has been reported recently that vagal fibers may convey sensory information from female pelvic organs to sensory nuclei in the brainstem (Komisaruk, Gerdes, & Whipple, 1997; Whipple, Richards, Tepper, & Komisaruk, 1995). The vagal pathway remains functional after spinal cord transection and may account for the menstrual cramping, analgesia, and orgasm reported in women with complete spinal cord transections.

Sensory innervation of the genitals is provided primarily by afferent fibers in the pudendal nerve. It has recently been reported that sensory fibers in the skin of the penis contain receptors for melanocortin (Van der Ploeg et al., 2002). Melanocortin is a peptide that has roles as a pituitary hormone and a neurotransmitter. It is being investigated for use in the treatment of sexual dysfunction (see below). Pudendal nerve afferents terminate in the medial portions of the dorsal horn and in the central gray matter (McKenna & Nadelhaft, 1986).

Interneurons

Interneurons relevant to sexual function have been located by neurophysiological and anatomical studies. Neurons activated by stimulation of the pudendal nerve or by perineal and pelvic visceral stimulation were located in the medial portions of the lumbosacral spinal gray (Fedirchuk, Song, Downie, & Shefchyk, 1992; Honda, 1985). Spinal interneurons related to pelvic function have been identified by a different technique. Strong activation of neurons often causes expression of the immediate early gene, c-fos, and its gene product Fos (Sagar, Sharp, & Curran, 1988). Stimulation of genital afferents resulted in labeled neurons in the medial dorsal horn, the central gray commissure, and the region of the intermediolateral cell column (a small nucleus at the lateral edge of the gray matter), consistent with the distribution of pelvic sensory terminals. The distribution of interneurons was similar in males and females (Birder, Roppolo, Iadarola, & de Groat, 1991; Lee & Erskine, 1996; Rampin, Bernabe, & Giuliano, 1997).

The use of viral tracing has been used to address the question of pelvic interneurons. Viruses, such as the pseudorabies virus, are picked up by nerve terminals near the injection site, retrogradely transported to the neuronal cell body, replicated, and picked up by nerve terminals presynaptic to the infected neurons (Card et al., 1993; Kuypers & Ugolini, 1990). Similar patterns of labeling were observed following injection into the penis, clitoris, uterus, and striated perineal muscles (Marson, 1995; Marson & McKenna, 1996; Marson, Platt, & McKenna, 1993; Papka et al., 1995; Papka, Williams, Miller, Copelin, & Puri, 1998; Tang, Rampin, Giuliano, & Ugolini, 1999). The majority of labeled neurons in the spinal cord were located in the central gray region of the spinal cord and near the intermediolateral cell column. These studies using different techniques all indicated that pelvic and sexual reflexes are dependent on spinal neurons in the central gray region of the lumbosacral segments.

The interneurons mediating ejaculation have been identified recently. Using neuroanatomical and behavioral techniques, a group of neurons in the central gray of spinal segments L3 and L4 was identified. When lesioned, the rats were unable to ejaculate, in copulatory testing, with no other sexual deficits (Truitt & Coolen, 2002). Note that these neurons were a discrete cluster of cells located within the larger pool of interneurons previously identified. Perhaps, additional pools of interneurons mediating other sexual and pelvic functions will be similarly identified.

Spinal Reflexes

Spinal reflexes, under supraspinal control, generate most sexual functions. Sexual reflexes are largely activated by pudendal afferents. One such reflex is the bulbocavernosus reflex. This reflex is a polysynaptic response seen in males and females elicited by light touch of the penis or clitoris. This stimulation of pudendal sensory fibers activates pudendal motoneurons to contract the striated perineal muscles (Bors & Blinn, 1959; McKenna & Nadelhaft, 1989; Rattner, Gerlaugh, Murphy, & Erdman, 1958; Vodusek, 1990). This reflex could be the basis for some sexual responses. For example, stimulation of the clitoris could lead to the development of the orgasmic platform (contraction of the circumvaginal muscles). In males, this reflex would also lead to contraction of the ischiocavernosus and bulbospongiosus muscles, which enhances penile erection. Stimulation of the clitoris and penis causes an inhibition of bladder activity by inhibition of pelvic nerve activity to the bladder and an increase in hypogastric nerve activity to the bladder neck. This inhibits the detrusor muscle of the bladder and contracts the bladder neck (Erlandson, Fall, & Carlsson, 1977; Lindstrom, Fall, Carlsson, & Erlandson, 1983). These reflexes strengthen urinary continence during sexual arousal and prevent retrograde ejaculation. Stimulation of the pudendal nerve sensory fibers also evokes long latency discharges in the cavernous nerve, indicating a polysynaptic reflex (Steers, Mallory, & de Groat, 1988). Stimulation of the pudendal penile sensory innervation also results in increases in intracavernous pressure (Bernabé, Rampin, Giuliano, & Benoit, 1995; Rampin et al. 1997. The mechanisms underlying female sexual responses remain to be elucidated directly.

Evidence indicates that ejaculation or sexual climax is also a spinal level reflex. Following complete spinal cord injury, women are still able to experience orgasm (Sipski, Alexander, & Rosen, 1985) and men are able to ejaculate (Bors & Comarr, 1960; Brindley, 1986b). In anesthetized, acutely spinalized male and female rats, genital stimulation gives rise to a response that resembles climax in humans: rhythmic contractions of the striated perineal muscles and vaginal and uterine contractions in females and penile erection and ejaculation in males (McKenna, Chung, & McVary, 1991). This climaxlike response also includes strong activation of the cavernous nerve, driven by both hypogastric and pelvic nerve preganglionic activity.

Medulla and Pons

A descending inhibitory control of spinal sexual reflexes has long been known (Beach, 1967). One site, nucleus paragigantocellularis (nPGi), has been identified in males as important in inhibitory control of spinal sexual reflexes. This brainstem nucleus is located in the ventral part of the rostral portion of the medulla. Neurons in the nPGi receive genital sensory information in males and females (Hubscher & Johnson, 1996; Rose, 1990). Axons from the nPGi project to pelvic efferent neurons and interneurons in the lumbosacral spinal cord (Marson & McKenna, 1990). Neurons in this area are transneuronally labeled following virus injection into the penis (Marson et al., 1993), the clitoris (Marson, 1995), and perineal muscles (Marson & McKenna, 1996; Tang et al., 1999). Transection of the spinal cord eliminates a tonic descending inhibition of spinal sexual reflexes. Lesions of this nucleus are as effective as spinal transection for eliminating the inhibition (Marson & McKenna, 1990). Chronic lesions of the nPGi facilitated ex copula penile reflexes (Marson, List, & McKenna, 1992). Ex copula erections are a type of erection seen in restrained, unanesthetized rats, in response to retraction of the prepuce of the penis. Male copulatory behavior was enhanced after chronic nPGi lesions, further indicating an inhibitory role for this nucleus (Yells, Hendricks, & Prendergast, 1992). Over 80% of the spinally projecting neurons in the nPGi stain positively for serotonin. Further, serotonin applied to the spinal cord inhibits spinal sexual reflexes (Marson & McKenna, 1992). The serotonergic inhibition of sexual reflexes by the nPGi is a possible substrate for the high incidence of orgasmic dysfunction seen with the use of selective serotonin reuptake inhibitor (SSRI) antidepressants, which elevate brain serotonin levels (Lane, 1997; Modell, Katholi, Modell, & DePalma, 1997). The serotonergic receptor subtype mediating the inhibition is unknown. Serotonergic projections to lumbosacral cord from the brainstem nuclei raphe pallidus, magnus, and parapyramidal region have been demonstrated (Du, 1989; G. Holstege, Kuypers, & Boer, 1969; J. C. Holstege, 1987; Loewy & McKellar, 1981; Martin, Vertes, & Waltzer, 1985). The functional roles of these serotonergic pathways are not known. The purpose of the tonic inhibition of sexual reflexes is unknown. It may be suggested that it serves to suppress sexual responses to incidental genital stimuli.

A number of other brainstem nuclei have been shown to project to relevant areas in the lumbosacral spinal cord, but their functional role in sexual response is unknown. There are significant noradrenergic projections from the A5 catecholaminergic cell group and from locus ceruleus (Loewy, McKellar, & Saper, 1979; Nygren & Olson, 1977). These provide a dense innervation of pudendal motoneurons, sympathetic and parasympathetic preganglionic neurons, and interneurons (Kojima et al., 1985; Schrøder & Skagerberg, 1985). These noradrenergic nuclei are also consistently labeled following pseudorabies virus injection into sexual organs (Marson, 1995; Marson & McKenna, 1996; Marson et al., 1993; Tang et al., 1999).

Barrington's nucleus in the parabrachial complex of the pons, also called the *pontine micturition center*, plays a key role in the descending control of bladder and sphincter reflexes (Barrington, 1925; De Groat & Steers, 1990; G. Holstege, Griffiths, De Wall, & Dalm, 1986; Kuru, 1965). In addition, Barrington's nucleus has also been shown to be involved in defecation reflexes and straining responses related to defecation and parturition (Fukuda & Fukai, 1986a, 1986b, 1988; Fukuda, Fukai, Yamane, & Okada, 1981). This nucleus has also been transneuronally labeled following injection of viral tracers into the penis, clitoris, and the ischiocavernosus and bulbospongiosus muscles (Marson, 1995; Marson & McKenna, 1996; Marson et al., 1993). These findings suggest that Barrington's nucleus has an important role in the control of several pelvic functions, including sexual function.

Midbrain

In the midbrain, the periaqueductal gray is known to be an important area for the control of homeostatic functions and motivated behaviors, including sexual function. It has extensive connections with all of the brainstem sites just discussed, including the nPGi (Bandler & Shipley, 1994; Van Bockstaele, Pieribone, & Aston-Jones, 1989), and has connections with many hypothalamic sites involved in sexual function (Behbehani, 1995; Simerly & Swanson, 1986, 1988). A high concentration of androgen receptors has also been identified in this region (Gréco, Edwards, Michael, & Clancy, 1996). Stimulation and recording studies and c-fos experiments have provided evidence that the ventral tegmental field of the midbrain has an important role in transmitting genital sensory signals to diencephalic structures (Baum & Everitt, 1992; Rose, 1990; Shimura & Shimokochi, 1991). Neurons in this midbrain are labeled following viral injection into the penis, clitoris, uterus, and perineal muscles (Marson, 1995; Marson & McKenna, 1996; Marson et al., 1993; Papka et al., 1998; Tang et al., 1999). Midbrain lesions have been shown to produce major sexual deficits (Barfield, Wilson, & McDonald, 1975; Brackett & Edwards, 1984; Brackett, Iuvone, & Edwards, 1986; T. K. Clark, Caggiula, McConnel, & Antelman, 1975).

Hypothalamus

The medial preoptic area (MPOA) and adjacent regions of the hypo-thalamus has long been considered necessary for the control of sexual behavior. Lesions of neurons in the MPOA disrupt copulatory behavior in every vertebrate species examined to date (Hart & Leedy, 1985; Meisel & Sachs, 1994; Rose, 1990). Male sexual behavior is activated by electrical or chemical stimulation of this area in conscious animals (Davidson, 1966; Malsbury, 1971; Merari & Ginton, 1975) and can elicit sexual responses in anesthetized animals (Giuliano et al., 1996; Marson & McKenna, 1994). Neurons in the MPOA contain androgen receptors (Sar & Stumpf, 1975), and implantation of testosterone into the MPOA restores sexual behavior in castrated animals (Davidson, 1966). The MPOA has extensive connections with widespread areas of the limbic system and brainstem (Simerly & Swanson, 1986, 1988).

Following lesions of the MPOA, male copulatory behavior is severely impaired in animals. Despite this, the animals retain their capacity to have erection and ejaculation, and they retain their sexual motivation. In an assessment of sexual motivation using a paradigm in which rats pressed a bar to gain access to estrous female rats, the MPOA rats continued to be motivated to have access to the female rats (Everitt & Stacey, 1987). Following lesions of the MPOA in monkeys, copulatory behavior was severely impaired. However, lesioned male monkeys also continued to bar press for female monkeys and continued to masturbate to ejaculation (Slimp, Hart, & Goy, 1975). Thus, it appears that the MPOA does not generate sexual behavior but is important for integrating sensory and hormonal cues related to sexual behavior.

The paraventricular nucleus (PVN) of the hypothalamus appears to be a crucial site for the control of sexual function. Neurons in this area are activated by genital sensory stimulation (Yanagimoto, Honda, Goto, & Negoro, 1996). In turn, neurons in the parvocellular division of the PVN project to multiple levels of the spinal cord, including direct projections to pelvic autonomic and somatic efferents (Cechetto & Saper, 1988; Saper, Loewy, Swanson, & Cowan, 1976; Wagner & Clemens, 1991). The PVN receives an extensive input from the MPOA (Simerly & Swanson, 1988). The lumbosacral projection is mediated in part by oxytocin neurons (Tang, Rampin, Calas, Facchinetti, & Giuliano, 1998; Veronneau-Longueville et al., 1999; Wagner & Clemens, 1993). The PVN was consistently labeled after pseudorabies virus injection into the penis, clitoris, and the penile muscles (Marson, 1995; Marson & McKenna, 1996; Marson et al., 1993). Stimulation of the PVN elicited seminal discharge in unanesthetized rats (Van Dis & Larsson, 1970) and erection in anesthetized rats (Chen, Chan,

Chang, & Chan, 1997). Lesions of the PVN caused disruption of seminal emission in copulatory testing in rats (Ackerman, Lange, & Clemens, 1997).

Forebrain

The medial portion of the amygdala is a crucial region for sexual behavior. Lesions of the amygdala give rise to disruption in sexual behavior, particularly aspects of sexual motivation (Everitt, Cador, & Robbins, 1989). The hypersexuality induced by the large lesions reported by Kluver and Bucy (1939) are likely due to the removal of inhibitory control by pyriform cortex that is destroyed in such lesions and not due to damage to the amygdala. Neurons in the medial amygdala are labeled with c-fos following copulatory behavior in both males and females (Baum & Everitt, 1992; Coolen, Peters, & Veening, 1996; Tetel, Getzinger, & Blaustein, 1993).

Forebrain regions involved in female sexual function have largely been identified on the basis of Fos staining in copulatory tests and viral staining. Medial amygdala, bed nucleus of the stria terminalis, and some other regions are most consistently identified (Coolen et al., 1996; Erskine, 1993; Papka et al., 1998; Tetel et al., 1993; Veening & Coolen, 1998). The Fos labeling in these regions is strongly affected by vaginocervical stimulation during copulation.

PHARMACOLOGY OF SEXUAL FUNCTION

Because of the extensive role of the central nervous system, and therefore of neurons, in sexual activity, neurological pharmacological agents have a strong effect on sexual function. This effect varies from drug to drug, and sometimes from species to species. In the rest of the chapter, I elucidate the studies of these various drugs and their effects on sexual function.

Serotonin

There is an extensive serotonergic innervation of most of the brain and spinal sites involved in the control of sexual function (Bancila et al., 1999; Loewy & McKellar, 1981; Marson & McKenna, 1992; Skagerberg & Bjorklund, 1985; Steinbusch, 1981; Tang et al., 1998). It is generally agreed that in large part, serotonin exerts an inhibitory effect on sexual function (Bitran & Hull, 1987). Decreased serotonin levels, by inhibition of its synthesis or by lesions of serotonergic neurons, resulted in enhancement of sexual function (Kondo, Yamanouchi, & Arai, 1993; Marson et al., 1992; McIntosh & Barfield, 1984; Yells et al., 1992). In contrast, administration of serotonin or drugs that increase serotonin levels inhibited sexual function (Ahlenius et al., 1981; Marson & McKenna, 1992; Svensson & Hansen, 1984; Szele, Murphy, & Garrick, 1988). The major effect of serotonin appears to be an inhibition of ejaculation or sexual climax. This is the primary sexual side effect seen with SSRI antidepressants, which increase serotonin levels (Lane, 1997; Modell et al., 1997). An inhibition of ejaculation in copulatory testing in rats has been noted (Svensson & Hansen, 1984), and climaxlike sexual reflexes were inhibited by serotonin (Marson & McKenna, 1992).

Thirteen serotonin receptor subtypes have been identified (Barnes & Sharp, 1999; Gerhardt & van Heerikhuizen, 1997). The receptor subtypes on the various neurons involved in the control of sexual function have not been identified. The 1A, 1B, 2A, and 2C receptor subtypes have been found in the spinal cord (Marlier, Teilhac, Cerruti, & Privat, 1991; Ridet, Tamir, & Privat, 1994; Thor, Nickolaus, & Helke, 1993). The 2C receptor has been identified with neuroanatomical techniques on lumbosacral neurons specifically related to the control of penile erection (Bancila et al., 1999). The functional role of these receptors is not entirely clear. Drugs, which have actions at the 2C receptor, have been found to facilitate erection in rats. 1-(3-Chlorophenyl)-piperazine (m-CPP) and N-trifluoromethylphenyl-piperazine (TFMPP) are partial agonists at 2C receptors as well as 2A receptor antagonistic actions (Barnes & Sharp, 1999). These induce erection and firing in the cavernous nerve; however, they also inhibit ejaculation and some aspects of sexual behavior (Aloi, Insel, Mueller, & Murphy, 1984; Berendsen & Broekkamp, 1987; Berendsen, Broekkamp, & van Delft, 1991; Berendsen, Jenck, & Broekkamp, 1990; Millan, Peglion, Lavielle, & Perrin-Monneyron, 1997; Pomerantz, Hepner, & Wertz, 1993; Steers & de Groat, 1989; Szele et al., 1988). Another 2C agonist, RSD 992, induced erections (Hayes, Doherty, Hanson, Gorzalka, & Adaikan, 2000). These data suggest that 2C receptors in the spinal cord may facilitate erections. It has also been hypothesized that 2C receptors may be inhibitory to ejaculation (Waldinger, Berendsen, Blok, Olivier, & Holstege, 1998).

A serotonin 1A receptor agonist, 8-hydroxy-2-(di-n-propylamino) tetralin (8-OH DPAT), has been studied extensively for its effects in sexual function (Bitran & Hull, 1987). This drug has complex effects on sexual behavior. Generally, it has inhibitory effects on erection but greatly facilitates ejaculation (Ahlenius et al., 1981; Fernandez-Guasti, Escalante, Ahlenius, Hillegaart, & Larsson, 1992). This drug also facilitated a climaxlike reflex in anesthetized rats (Carro-Juarez & Rodriguez-Manzo, 2001). However, the specificity of the action of this drug is in question, because its effects on male rat sexual behavior are blocked by antagonists of dopamine D2 receptors and not serotonergic 1A antagonists (Matuszewich et al., 1999). However, a more selective 1A agonist also reduced ejaculatory latency (Foreman et al., 1994). It is clear that further research is needed to clarify the role of serotonin 1A receptors in sexual function.

Trazodone is an antidepressant that is pharmacologically different from the SSRIs. Its major metabolite, m-CCP, is an agonist at serotonin 2C receptors and antagonist at 2A receptors (Barnes & Sharp, 1999; Monsma, Shen, & Ward, 1993). mCPP induces erection in rats and increases the neural activity of the cavernous nerve (Steers & de Groat, 1989). The mode of action of trazodone in depression is not fully understood; it has a marked sedative action. Trazodone has been associated with cases of priapism (prolonged, painful erection) in men (Azadzoi, Payton, Krane, & Goldstein, 1990) and clitoral priapism in women (Pescatori et al., 1993). It has been shown to increase nocturnal tumescence in men (Saenz de Tejada et al., 1991). The site and mechanism of its sexual side effects are not known.

Adverse sexual side effects are commonly seen with SSRI antidepressants (Lane, 1997; Modell et al., 1997). The most common effect is to delay or inhibit ejaculation in men and orgasm in women. These drugs increase serotonin levels in the brain; however, their pharmacology is complex, and many neurotransmitter systems are affected by them. This effect of SSRIs is now being examined for their use in the treatment of premature ejaculation (Waldinger, 2002). The rationale is that at an appropriate dose, ejaculation can be delayed but not totally inhibited in these patients.

Dopamine

The medial preoptic area and PVN are extensively innervated by dopaminergic neurons of the incertohypothalamic pathway (Bjorklund, Lindvall, & Nobin, 1975). There is also a dopaminergic innervation of the lumbosacral spinal cord from diencephalic dopaminergic cells (Skagerberg, Bkorklund, Lindvall, & Schmidt, 1982; Skagerberg & Lindvall, 1985). Thus, dopamine is anatomically organized to modulate the expression of sexual function. Numerous pharmacological studies in animals and humans support this conclusion. In general, dopaminergic agonists facilitate several aspects of sexual function, from erection, ejaculation, female sexual arousal, and sexual motivation.

The sexual effects of dopamine seem to involve both the dopaminergic D1 and D2 receptors, acting in the medial preoptic area and the PVN (Hull, 1995). In the MPOA of male rats, there was a consistent rise in dopamine release prior to copulation (Hull, Du, Lorrain, & Matuszewich, 1995). This release was anatomically restricted and specific to sexual stimuli. The release of dopamine in the MPOA can be mimicked by microinjections of dopaminergic agonists. Microinjection of apomorphine, a nonselective D1/D2 receptor agonist, increased the rate and efficiency of copulation and facilitated erections (Bazzett et al., 1991; Hull et al., 1986; Hull et al., 1992). Injection of dopaminergic antagonists into the MPOA decreased the number of penile reflexes (Pehek, Thompson, & Hull, 1988a; Warner et al., 1991). Injection of apomorphine into the PVN, but not the MPOA, enhanced seminal emission. Results have been presented that indicate that D1 receptor stimulation increases erections and D2 stimulation promotes ejaculation (Hull, 1995). In contrast, following microinjection in the PVN, it appears that D2 rather than D1 receptors primarily facilitate erections (Melis, Argiolas, & Gessa, 1987). The proerectile effect of dopamine injected into the PVN was enhanced by administration of selegiline, a drug that inhibits the degradation of dopamine (Allard et al., 2002). These results indicate that dopaminergic stimulation of the hypothalamus facilitates male sexual responses. Note, however, that contrary to these results in rats, rhesus monkeys did not respond to apomorphine, indicating possible species differences in the dopaminergic control of sexual function (Chambers & Phoenix, 1989).

In the male, the excitatory effect of paraventricular dopaminergic stimulation is mediated in part by oxytocin. Oxytocinergic cell bodies in the PVN receive dopaminergic innervation (Buijs, 1978; Lindvall, Björklund, & Skagerberg, 1984). Penile erection induced by apomorphine can be blocked by oxytocin receptor antagonists (Argiolas, Melis, Vargiu, & Gessa, 1987; Melis, Argiolas, & Gessa, 1989). In contrast, penile erections were induced by injection of oxytocin into the PVN, and these were not blocked by dopamine receptor antagonists (Melis et al., 1989). These results suggest that dopamine stimulates oxytocinergic neurons in the PVN, which then cause the erectile response.

Dopamine also appears to exert a facilitatory effect on sexual responses at the spinal level. Intrathecal injection of apomorphine at the lumbosacral facilitated ejaculation, although it suppressed ex copula erections and slowed the rate of copulation (Pehek, Thompson, & Hull, 1989a, 1989b). In anesthetized rats, intrathecal injection of apomorphine evoked erection in both normal and spinalized rats (Giuliano et al., 2002).

The role of dopamine in female sexual function is less clear. Dopamine in the MPOA is increased by estrogen and progesterone administration in ovariectomized female rats, which induces receptivity. Dopamine further increased when copulation began (Matuszewich, Lorrain, & Hull, 2000). Sexual receptivity, as measured by the lordosis quotient, was increased by administration of apomorphine or D2 agonists (Foreman & Hall, 1987; Hamburger-Bar & Rigter, 1975). Proceptive behavior was also increased by D2 agonists (Everitt, Fuxe, Hokfelt, & Jonsson, 1975). The role of dopamine on female genital arousal or sexual climax has not been investigated. In humans, data suggest that dopamine increases sexual motivation or desire. Dopaminergic agents used for the treatment of Parkinsonism sometimes induce increases in sexual desire (Courty, Durif, Zenut, Courty, & Lavarenne,

1997; Uitti et al., 1989). However, this may be a rare finding (Goodwin, 1971). There are anecdotal reports that cocaine and amphetamine, which increase dopamine levels, may be aphrodisiac. However, chronic use of stimulants is associated with severe sexual dysfunction, including anorgasmia and lack of sexual desire (Miller & Gold, 1988).

The effects of apomorphine on male sexual function in animals have led to the development of apomorphine as a treatment for erectile dysfunction in men. Apomorphine is able to induce erection in normal (Lal, Ackman, Thavundayil, Kiely, & Etienne, 1998) and impotent men (Lal et al., 1987; Segraves, Bari, Segraves, & Spirna, 1991). No effect of apomorphine was noted on libido (Lal et al., 1984; Julien & Over, 1984). However, apomorphine can cause significant side effects, including nausea, vomiting, drowsiness, flushing, and dizziness (Lal et al., 1984; Segraves et al., 1991). A sublingual preparation designed to minimize the side effects has been approved in Europe for the treatment of erectile dysfunction (Heaton, 2001).

Noradrenaline

Noradrenergic neurons from the A5 region, the locus ceruleus, and subceruleus project to the nuclei in the spinal cord involved in erection (Schrøder & Skagerberg, 1985). Generally, it appears that increased noradrenergic activity stimulates, and decreased noradrenergic activity inhibits, sexual function (Bitran & Hull, 1987). Hypothalamic nuclei are also targets of noradrenergic fibers from pontine nuclei, especially the locus ceruleus (Sawchenko & Swanson, 1982).

Both the alpha-1 adrenoceptor antagonist, prazosin, and the alpha-2 adrenoceptor agonist, clonidine, depressed copulation in male rats (J. T. Clark, Smith, & Davidson, 1985). The alpha-2 antagonist yohimbine increased sexual motivation and performance (J. T. Clark, Smith, & Davidson, 1984). Yohimbine also restored sexual activity in sexually exhausted male rats (Rodriguez-Manzo & Fernandez-Guasti, 1994). Clonidine delivered intracerebroventricularly or into the MPOA suppressed copulatory behavior (J. T. Clark, 1991). Lesion of the locus ceruleus or administration of noradrenaline synthesis inhibitors inhibited sexual function by increasing the postejaculatory refractory period and the mount and intromission latencies (McIntosh & Barfield, 1984). An alpha-1 agonist was also shown to facilitate ejaculation (J. T. Clark, Kalra, & Kalra, 1987). Intrathecal noradrenaline accelerated pelvic thrusting (Hernandez-Gonzalez, Oropeza, Guevara, Cervantes, & Morali, 1994). The role of central noradrenaline in female sexual function is less understood. Noradrenergic nuclei in the brainstem are activated by copulatory behavior in female rats (Yang & Voogt, 2001). However, lesions of the noradrenergic innervation of the hypothalamus did not affect receptive behavior in female rats (Davis, Manzanares, Lookingland, Moore,

& Clemens, 1991). No studies have been reported on the effects of central noradrenergic pathways on female sexual genital arousal or climax.

The effects of noradrenergic agents in male animals have led to studies assessing their potential use in the treatment of erectile dysfunction in men. Phentolamine, an alpha-1 receptor antagonist, has been evaluated for its use in treating erectile dysfunction. It was found to be moderately effective (I. Goldstein, 2000; I. Goldstein, Carson, Rosen, & Islam, 2001; Gwinup, 1988; Zorgniotti, 1992, 1994; Zorgnotti & Lizza, 1994). A direct comparison with sildenafil (Viagra) or one of the newer phosphodiesterase Type 5 inhibitors will be necessary to determine it relative efficacy and side-effect profile. Yohimbine is an alkaloid extracted from an African bark. It is an alpha-2 antagonist that has been used for over a century for sexual purposes either as an herbal preparation or the pure compound (Morales, 2000). Its site of action is probably in the central nervous system, because intracavernous injection of alpha-2 antagonists does not produce erection (Brindley, 1986a). Trials with patients who experience organic impotence showed only marginal effects (Morales, Condra, & Owen, 1987). Slightly better results were seen with patients who experience psychogenic impotence (Morales et al., 1987; Reid et al., 1987). A meta-analysis of yohimbine trials demonstrated that yohimbine was superior to placebo in the treatment of erectile dysfunction. However, the overall effect is not remarkable (Carey & Johnson, 1996; Ernst & Pittler, 1998).

Oxytocin

Spinal autonomic nuclei are innervated by oxytocin containing neurons in the PVN (Sawchenko & Swanson, 1982; Swanson & Kuypers, 1980; Wagner & Clemens, 1991). The PVN was consistently labeled after pseudorabies virus injection into the penis and the penile muscles, ischiocavernosus and bulbospongiosus (Marson & McKenna, 1996; Marson et al., 1993; Tang et al., 1999). The lumbosacral projections to the lumbosacral spinal cord neurons, identified as projecting to the pelvic organs, were found to contain oxytocin (Tang et al., 1999). Oxytocin receptors were found in the sacral parasympathetic nucleus and interneurons in the dorsal gray commissure of the L6-S1 spinal cord (Veronneau-Longueville et al., 1999). The PVN is activated during sexual arousal, and especially at orgasm, as indicated by the fact that plasma oxytocin concentrations increased in male and female humans (Carmichael et al., 1987; Murphy, Seckl, Burton, Checkley, & Lightman, 1987).

Oxytocin activates penile erections when injected into the lateral ventricles or directly into the PVN in rats (Argiolas, 1992; Argiolas, Melis, & Gressa, 1986; Melis, Succu, Iannucci, & Argiolas, 1997a). The explanation for the effect of oxytocin injected into the PVN is that oxytocin stimulates oxytocinergic receptors located on the cell bodies of oxytocinergic neurons in the PVN (Argiolas, 1992; Argiolas et al., 1986). Thus, oxytocin neurons are excitatory to themselves. An oxytocin innervation of oxytocinergic neurons has been identified in the supraoptic and PVN nuclei (Theodosis, 1985).

A spinal site of action of oxytocin has also been identified. Intrathecal administration of oxytocin and a specific oxytocin agonist were found to cause erection in a dose-dependent manner in anesthetized rats (Giuliano, Bernabe, McKenna, Longueville, & Rampin, 2001). The erectile response was blocked by a specific oxytocin antagonist and was not elicited by a closely related peptide, vasopressin. These results indicate that oxytocin may be proerectile at both hypothalamic and spinal sites. The effects of oxytocin on ejaculation and female sexual function have not been evaluated. However, there is a considerable literature on the role of oxytocin in pair bonding (Carter, 1998).

Adrenocorticotropin and Related Peptides

Several biologically active peptides are derived from pro-opiomelanocortin, including adrenocorticotropic (ACTH) and the alpha-melanocyte-stimulating hormones (alpha-MSH). Both ACTH and alpha-MSH have been shown to give rise to erectile responses. A syndrome of penile erection and ejaculation, grooming, stretching, and yawning is produced after injection into the brain or hypothalamus (Argiolas, Melis, Murgia, & Schioth, 2000; Bertolini, Gessa, & Ferrari, 1975; Ferrari, Gessa, & Vargiu, 1963; Mains, Eippers, & Ling, 1977; Poggioli, Arletti, Benelli, Cavazzuti, & Bertolini, 1998). A major site of action is in the PVN, dorsomedial nucleus, ventromedial nucleus, and anterior hypothalamic area (Argiolas et al., 2000). However, there also appears to be a spinal, and possibly peripheral, site of action. A synthetic melanocortin peptide, Melanotan II, induced erection in spinal transected rats (Wessels et al., 2003). The sexual effects of alpha-MSH and ACTH are due to stimulation of the melanocortin (MC) receptors. There are five different melanocortin receptor subtypes (Wikberg, 1999; Wikberg et al., 2000). There are several major drug development programs related to these receptors because melanocortin inhibits feeding, and drugs stimulating these receptors may be useful in treating obesity. The proerectile effect appears to be largely due to stimulation of MC4 receptors (Van der Ploeg et al., 2002). It is interesting that these authors also reported MC4 receptors on sensory receptors in the glans of the penis. This suggests that melanocortins may also modulate genital sensory stimulation. Melanotan II has been administered subcutaneously to men with erectile dysfunction. It induced potent erections (Wessels et al., 1998; Wessels, Levine, Hadley, Dorr, & Hruby, 2000). A closely related compound, PT 141, is being tested as a treatment for erectile dysfunction using intranasal delivery (peptides are degraded when given orally).

Opiates

Chronic use of opiates is associated with severe sexual dysfunction, suggesting a central regulation of sexual function by endogenous opiate systems (Crowley & Simpson, 1978; Cushman, 1972). Administration of opiates depresses copulatory behavior in rats injected systemically (McIntosh, Vallano, & Barfield, 1980; Pfaus & Gorzalka, 1987) into the brain or hypothalamus (Hughes, Everitt, & Herbert, 1987; McIntosh et al., 1980; Melis, Stancampiano, Gessa, & Argiolas, 1992; Melis, Succu, & Argiolas, 1997). The sexual inhibitory effects of opiates are believed to be due to the stimulation of mu opiate receptor on oxytocinergic neurons in the PVN (Melis, Succu, Iannucci, & Argiolas, 1997b). In anesthetized animals, injection of the opiate antagonist naloxone induced erections (Domer, Wessler, Brown, & Matthews, 1988) and potentiated the erectile response of apomorphine in unanesthetized animals (Berendsen & Gower, 1986). In normal humans, naloxone was without effect on arousal (A. Goldstein & Hansteen, 1977). Another opiate antagonist that is orally active, naltrexone, was administered to impotent men in two small trials. It caused a significant improvement in erectile function (Fabbri et al., 1989; J. A. Goldstein, 1986).

Nitric Oxide

The role of nitric oxide in mediating the vasodilation during penile erection is well known. Anatomical and pharmacological evidence indicate that nitric oxide may be part of the central nervous system control of sexual function. There is a high concentration of nitric oxide synthase (NOS), the synthetic enzyme for nitric oxide, in the PVN (Bredt, Hwang, & Snyder, 1990). The PVN also contains a splice variant isoform of NOS that is found primarily in the penis (Ferrini, Magee, Vernet, Rajfer, & Gonzalez-Cadavid, 2003).

A crucial role for nitric oxide in the PVN has been identified. Inhibition of nitric oxide in the PVN inhibits the proerectile effects of dopamine and oxytocin (Argiolas, 1994). Nitric oxide donors injected into the PVN cause erection (Melis & Argiolas, 1995). Levels of nitric oxide in the PVN increase during erection and copulation (Melis, Succu, Iannucci, & Argiolas, 1998). Nitric oxide also plays a role in the MPOA. Nitric oxide levels there increase during copulation, and nitric oxide inhibition decreased copulatory behavior (Sato et al., 1999; Sato, Horita, Kurohata, Adachi, & Tsukamoto, 1998). Anatomical studies have demonstrated localization of nitric oxide synthase in neurons involved in the control of pelvic and autonomic function (Burnett et al., 1995; Dun et al., 1993; Saito et al., 1994; Valtschanoff, Weinberg, & Rustioni, 1992). Intrathecal administration of nitric oxide inhibitor blocked the proerectile effects of intrathecally delivered oxytocin, indicating that the effects of oxytocin in the spinal cord are mediated through nitric oxide interneurons. Volatile nitrites ("poppers") have been used extensively in sexual contexts. These produce an intense, brief "rush" with pounding heart, flushing, and mental dissociation. In sexual situations, users report heightened sexual arousal and loss of inhibitions. It is commonly reported that they relax smooth muscle of the anal sphincters, facilitating anal sex. However, whether this is for central nervous system effects or peripheral effects is unclear.

CONCLUSIONS

This chapter has demonstrated that there has been a tremendous increase recently of our understanding of the neurophysiology, neuroanatomy, and pharmacology of the central nervous system control of sexual function. The crucial role of the PVN and the nPGi has been identified. The inhibitory role of serotonin and the excitatory role of oxytocin and dopamine have been established. However, there exist many substantial gaps in our knowledge. Most of the research has been on the central control of penile erection. The central control and neurochemistry of ejaculation and sexual motivation are far less well understood. And, most glaringly, the central control of female sexual function is largely obscure. This is largely a result of issues relating to animal models. Most research has been focused on the neural and hormonal control of lordosis and female receptivity and proceptivity. Animal models for female genital responses are not as well developed as those for erection and ejaculation. Future research needs to be directed at these crucial issues.

REFERENCES

Ackerman, A. E., Lange, G. M., & Clemens, L. G. (1997). Effects of paraventricular lesions on sex behavior and seminal emission in male rats. *Physiology and Behavior, 63*, 49–53.

Adler, N. T., Davis, P. G., & Komisaruk, B. R. (1977). Variation in the size and sensitivity of a genital sensory field in relation to the estrous cycle in rats. *Hormones and Behavior, 9*, 334–344.

Ahlenius, S., Larsson, K., Svensson, L., Hjorth, S., Carlsson, A., & Lindberg, P. (1981). Effects of a new type of 5-HT receptor agonist on male rat sexual behaviour. *Pharmacology Biochemistry and Behavior, 15*, 785.

Allard, J., Bernabe, J., Derdinger, F., Alexandre, L., McKenna, K., & Giuliano, F. (2002). Selegiline enhances erectile activity induced by dopamine injection in the paraventricular nucleus of the hypothalamus in anesthetized rats. *International Journal of Impotence Research, 14*, 518–522.

Aloi, J. A., Insel, T. R., Mueller, E. A., & Murphy, D. L. (1984). Neuroendocrine and behavioral effects of m-chlorophenylpiperazine administration in rhesus monkeys. *Life Science, 34*, 1325–1331.

Andersson, K.-E., & Wagner, G. (1995). The physiology of penile erection. *Physiological Reviews, 75*, 191–236.

Argiolas, A. (1992). Oxytocin stimulation of penile erection: Pharmacology, site, and mechanism of action. *Annals of the New York Academy of Sciences, 652*, 194–203.

Argiolas, A. (1994). Nitric oxide is a central mediator of penile erection. *Neuropharmacology, 33*, 1339–1344.

Argiolas, A., Melis, M. R., & Gessa, G. L. (1986). Oxytocin: An extremely potent inducer of penile erection and yawning in male rats. *European Journal of Pharmacology, 130*, 265–272.

Argiolas, A., Melis, M. R., Murgia, S., & Schioth, H. B. (2000). ACTH- and alpha-MSH induced grooming, stretching, yawning and penile erection in male rats: Site of action in the brain and role of melanocortin receptors. *Brain Research Bulletin, 51*, 425–431.

Argiolas, A., Melis, M. R., Vargiu, L., & Gessa, G. L. (1987). d(CH2) 5Tyr (Me)-Orn8-vasotocin, a potent oxytocin antagonist, antagonizes penile erection and yawning induced by oxytocin and apomorphine, but not by ACTH (1–24). *European Journal of Pharmacology, 134*, 221–224.

Azadzoi, K. M., Payton, T., Krane, R. J., & Goldstein, I. (1990). Effects of intracavernosal trazodone hydrochloride: Animal and human studies. *Journal of Urology, 144*, 1277–1282.

Bancila, M., Vergé, D., Rampin, O., Backstrom, J. R., Sanders-Bush, E., McKenna, K. E., et al. (1999). 5-Hydroxytryptamine 2C receptors on spinal neurons controlling penile erection in the rat. *Neuroscience, 92*, 1523–1537.

Bandler, R., & Shipley, M. T. (1994). Columnar organization in the midbrain periaqueductal gray: Modules for emotional expression? *Trends in Neurosciences, 17*, 379–389.

Barfield, R. J., Wilson, C., & McDonald, P. G. (1975). Sexual behavior: Extreme reduction of postejaculatory refractory period by midbrain lesions in male rats. *Science, 189*, 147–149.

Barnes, N. M., & Sharp, T. (1999). A review of central serotonin receptors and their function. *Neuropharmacology, 38*, 1083–1152.

Barrington, F. J. F. (1925). The effects of lesions of the hind- and mid-brain on micturition in the cat. *Quarterly Journal of Experimental Physiology, 15*, 81–102.

Baum, M. J., & Everitt, B. J. (1992). Increased expression of c-fos in the medial preoptic area after mating in male rats: Role of afferent inputs from the medial amygdala and midbrain central tegmental field. *Neuroscience, 50*, 627–646.

Bazzett, T. J., Eaton, R. C., Thompson, J. T., Markowski, V. P., Lumley, L. A., & Hull, E. M. (1991). Dose dependent D2 effects on genital reflexes after MPOA injections of quinelorane and apomorphine. *Life Science, 48,* 2309–2315.

Beach, F. A. (1967). Cerebral and hormonal control of reflexive mechanisms involved in copulatory behavior. *Physiological Reviews, 47,* 289–316.

Behbehani, M. M. (1995). Functional characteristics of the midbrain periaqueductal gray. *Progress in Neurobiology, 46,* 574–605.

Bell, C. (1972). Autonomic nervous control of reproduction: Circulatory and other factors. *Pharmacological Reviews, 24,* 657–736.

Berendsen, H. H., & Broekkamp, C. L. (1987). Drug-induced penile erections in rats: Indications of serotonin 1B receptor mediation. *European Journal of Pharmacology, 135,* 279–287.

Berendsen, H. H., Broekkamp, C. L., & van Delft, A. M. (1991). Depletion of brain serotonin differently affects behaviors induced by 5HT1A, 5HT1C, and 5HT2 receptor activation in rats. *Behavior Neural Biology, 55,* 214–226.

Berendsen, H. H., & Gower, A. J. (1986). Opiate-androgen interactions in drug-induced yawning and penile erections in the rat. *Neuroendocrinology, 42,* 185–190.

Berendsen, H. H., Jenck, F., & Broekkamp, C. L. (1990). Involvement of serotonin 1C receptors in drug-induced penile erections in rats. *Psychopharmacology (Berlin), 101,* 57–561.

Berkley, K. J., Robbins, A., & Sato, Y. (1987). Uterine afferent fibers in the rat. In R. F. Schmidt, H.-G. Schaible, & C. Vahle-Hinz (Eds.), *Fine afferent nerve fibers and pain* (pp. 129–136). Weinheim, Germany: VCH Verlag.

Bernabé, J., Rampin, O., Giuliano, F., & Benoit, G. (1995). Intracavernous pressure changes during reflexive penile erections in the rat. *Physiology and Behavior, 57,* 837–841.

Bertolini, A., Gessa, G. L., & Ferrari, W. (1975). Penile erection and ejaculation: A central effect of ACTH-like peptides in mammals. In M. Sandler & G. L. Gessa (Eds.), *Sexual behavior: Pharmacology and biochemistry* (pp. 247–257). New York: Raven Press.

Birder, L. A., Roppolo, J. R., Iadarola, M. J., & de Groat, W. C. (1991). Electrical stimulation of visceral afferent pathways in the pelvic nerve increases c-fos in the rat lumbosacral spinal cord. *Neuroscience Letters, 129,* 193–196.

Bitran, D., & Hull, E. M. (1987). Pharmacological analysis of male rat sexual behavior. *Neuroscience and Biobehavioral Review, 11,* 365–389.

Björklund, A., Lindvall, O., & Nobin, A. (1975). Evidence of an incertohypothalamic dopamine neuron system in the rat. *Brain Research, 89,* 29–42.

Bors, E., & Blinn, K. A. (1959). Bulbocavernosus reflex. *Journal of Urology, 82,* 128–130.

Bors, E., & Comarr, A. E. (1960). Neurological disturbances of sexual function with special reference to 529 patients with spinal cord injury. *Urological Survey, 10,* 191–222.

Brackett, N. L., & Edwards, D. A. (1984). Medial preoptic connections with the midbrain tegmentum are essential for male sexual behavior. *Physiology and Behavior, 32,* 79–84.

Brackett, N. L., Iuvone, P. M., & Edwards, D. A. (1986). Midbrain lesions, dopamine and male sexual behavior. *Behavioral Brain Research, 20,* 231–240.

Bredt, D. S., Hwang, P. M., & Snyder, S. H. (1990). Localization of nitric oxide synthase indicating a neural role for nitric oxide. *Nature (London), 347,* 768–770.

Breedlove, S. M., & Arnold, A. P. (1981). Sexually dimorphic motor nucleus in the rat lumbar spinal cord: Response to adult hormone manipulation, absence in androgen insensitive rats. *Brain Research, 225,* 297–307.

Brindley, G. S. (1986a). Pilot experiments on the actions of drugs injected into the human corpus cavernosum penis. *British Journal of Pharmacology, 87,* 495–500.

Brindley, G. S. (1986b). Sexual and reproductive problems of paraplegic men. In J. R. Clarke (Ed.), *Oxford reviews of reproductive biology* (pp. 214–222). Oxford, England: Clarendon Press.

Buijs, R. M. (1978). Intra- and extrahypothalamic vasopressin and oxytocin pathways in the rat: Pathways to the limbic system, medulla oblongata and spinal cord. *Cell and Tissue Research, 192,* 423–435.

Burnett, A. L., Lowenstein, C. J., Bredt, D. S., Chang, T. S. H., & Snyder, S. H. (1992). Nitric oxide: A physiologic mediator of penile erection. *Science, 257,* 401–403.

Burnett, A. L., Saito, S., Maguire, M. P., Yamaguchi, H., Chang, T. S., & Hanley, D. F. (1995). Localization of nitric oxide synthase in spinal nuclei innervating pelvic ganglia. *Journal of Urology, 153,* 212–217.

Card, J. P., Rinaman, L., Lynn, R. B., Lee, B.-H., Meade, R. P., Miselis, R. R., & Enquist, L. W. (1993). Pseudorabies virus infection of the rat central nervous system: Ultrastructural characterization of viral replication, transport, and pathogenesis. *Journal of Neuroscience, 13,* 2515–2539.

Carey, M. P., & Johnson, B. T. (1996). Effectiveness of yohimbine in the treatment of erectile disorder: Four meta-analytic integrations. *Archives of Sexual Behavior, 25,* 341–360.

Carlson, R. R., & De Feo, V. J. (1965). Role of the pelvic nerve vs. the abdominal sympathetic nerves in the reproductive function of the female rat. *Endocrinology, 77,* 1014–1022.

Carmichael, M. S., Humbert, R., Dixon, J., Palmisano, G., Greenleaf, W., & Davidson, J. M. (1987). Plasma oxytocin increases in the human sexual response. *Journal of Clinical Endocrinology and Metabolism, 64,* 27–31.

Carro-Juarez, M., & Rodriguez-Manzo, G. (2001). Exhaustion of the coital reflex in spinal male rats is reversed by the serotonergic agonist 8-OH-DPAT. *Behavioral Brain Research, 118,* 161–168.

Carter, C. S. (1998). Neuroendocrine perspectives on social attachment and love. *Psychoneuroendocrinology, 23,* 779–818.

Cechetto, D. F., & Saper, C. B. (1988). Neurochemical organization of the hypothalamic projection to the spinal cord in the rat. *Journal of Comparative Neurology, 272,* 579–604.

Chambers, K. C., & Phoenix, C. H. (1989). Apomorphine, deprenyl, and yohimbine fail to increase sexual behavior in rhesus males. *Behavioral Neuroscience, 103,* 816–823.

Chen, K. K., Chan, S. H. H., Chang, L. S., & Chan, J. Y. H. (1997). Participation of paraventricular nucleus of hypothalamus in central regulation of penile erection in the rat. *Journal of Urology, 158,* 238–244.

Clark, J. T. (1991). Suppression of copulatory behavior in male rats following central administration of clonidine. *Neuropharmacology, 130,* 373–382.

Clark, J. T., Kalra, S. P., & Kalra, P. S. (1987). Effects of a selective alpha 1-adrenoceptor agonist, methoxamine, on sexual behavior and penile reflexes. *Physiology and Behavior, 140,* 747–753.

Clark, J. T., Smith, E. R., & Davidson, J. M. (1984). Enhancement of sexual motivation in male rats by yohimbine. *Science, 1225,* 847–849.

Clark, J. T., Smith, E. R., & Davidson, J. M. (1985). Evidence for the modulation of sexual behavior by alpha-adrenoceptors in male rats. *Neuroendocrinology, 41,* 36–43.

Clark, T. K., Caggiula, A. R., McConnel, R. A., & Antelman, S. M. (1975). Sexual inhibition is reduced by rostral midbrain lesions in the male rat. *Science, 190,* 169–171.

Coolen, L. M., Peters, H. J. P. W., & Veening, J. G. (1996). Fos immunoreactivity in the rat brain following consummatory elements of sexual behavior: A sex comparison. *Brain Research, 738,* 67–82.

Courty, E., Durif, F., Zenut, M., Courty, P., & Lavarenne, J. (1997). Psychiatric and sexual disorders induced by apomorphine in Parkinson's disease. *Clinical Neuropharmacology 20,* 140–147.

Crowley, T. J., & Simpson, A. (1978). Methadone dose and human sexual behavior. *International Journal of Addiction, 31,* 285–295.

Cushman, P. (1972). Sexual behavior in heroin addiction and methadone maintenance. Correlation with plasma luteinizing hormone. *New York State Journal of Medicine, 72,* 1261–1265.

Davidson, J. M. (1966). Activation of male rat's sexual behavior by intracerebral implantation of androgen. *Endocrinology, 79,* 783–794.

Davis, B. L., Manzanares, J., Lookingland, K. J., Moore, K. E., & Clemens, L. G. (1991). Noradrenergic innervation to the VMN or MPN is not necessary for lordosis. *Pharmacology Biochemistry and Behavior, 39,* 737–742.

De Groat, W. C., & Booth, A. M. (1993). Neural control of penile erection. In C. A. Maggi (Ed.), *The autonomic nervous system: Nervous control of the urogenital system* (pp. 465–513). London: Harwood.

De Groat, W. C., & Steers, W. D. (1990). Autonomic regulation of the urinary bladder and sexual organs. In A. D. Loewy & K. M. Spyer (Eds.), *Central regulation of autonomic functions* (pp. 310–333). New York: Oxford University Press.

Domer, F. R., Wessler, G., Brown, R. L., & Matthews, A. (1988). Effects of naloxone on penile erection in cats. *Pharmacology Biochemistry and Behavior, 30*, 543–545.

Du, H.-J. (1989). Medullary neurons with projections to lamina X of the rat as demonstrated by retrograde labeling after HRP microelectrophoresis *Brain Research, 505*, 135–140.

Dun, N. J., Dun, S. L., Wu, S. Y., Forstermann, U., Schmidt, H. H., & Tseng, L. F. (1993). Nitric oxide synthase immunoreactivity in the rat, mouse, cat and squirrel monkey spinal cord. *Neuroscience, 54*, 845–857.

Erlandson, B. E., Fall, M., & Carlsson, C. A. (1977). The effect of intravaginal electrical stimulation on the feline urethra and urinary bladder: Electrical parameters. *Scandinavian Journal of Urology Nephrology Supplement, 44*, 5–18.

Ernst, E., & Pittler, M. H. (1998). Yohimbine for erectile dysfunction: A systematic review and meta-analysis of randomized clinical trials. *Journal of Urology, 159*, 433–436.

Erskine, M. S. (1993). Mating-induced increases in FOS protein in preoptic area and medial amygdala of cycling female rats. *Brain Research Bulletin, 32*, 447–451.

Everitt, B. J., Cador, M., & Robbins, T. W. (1989). Interactions between the amygdala and ventral striatum in stimulus–reward associations: Studies using a second-order schedule of sexual reinforcement. *Neuroscience, 30*, 63–75.

Everitt, B. J., Fuxe, K., Hokfelt, F. T., & Jonsson, C. (1975). Role of monoamines in the control by hormones of sexual receptivity in the female rat. *Journal of Comparative Physiology and Psychology, 89*, 556–572.

Everitt, B. J., & Stacey, P. (1987). Studies of instrumental behavior with sexual reinforcement in male rats: II: Effects of preoptic lesions, castration and testosterone. *Journal of Comparative Psychology, 101*, 407–419.

Fabbri, A., Jannini, E. A., Gnessi, L., Moretti, C., Ulisse, S., Franzese, A., et al. (1989). Endorphins in male impotence: Evidence for naltrexone stimulation of erectile activity in patient therapy. *Psychoneuroendocrinology, 14*, 103–111.

Fedirchuk, B., Song, L., Downie, J. W., & Shefchyk, S. J. (1992). Spinal distribution of extracellular field potentials generated by electrical stimulation of pudendal and perineal afferents in the cat. *Experimental Brain Research, 89*, 517–520.

Fernandez-Guasti, A., Escalante, A. L., Ahlenius, S., Hillegaart V., & Larsson, K. (1992). Stimulation of 5-HT1A and 5-HT1B receptors in brain regions and its effects on male rat sexual behavior. *European Journal of Pharmacology, 210*, 121–129.

Ferrari, W., Gessa, G. L., & Vargiu, L. (1963). Behavioural effects induced by intracisternally injected ACTH and MSH. *Annals of the New York Academy of Sciences, 104*, 330–345.

Ferrini, M. G., Magee, T. R., Vernet, D., Rajfer, J., & Gonzalez-Cadavid, N. F. (2003). Penile neuronal nitric oxide synthase and its regulatory proteins are present in hypothalamic and spinal cord regions involved in the control of penile erection. *Journal of Comparative Neurology, 458,* 46–61.

Foreman, M. M., Fuller, R. W., Rasmussen, K., Nelson, D. L., Calligaro, D. O., Zhang, L., et al. (1994). Pharmacological characterization of LY293284: A 5-HT1A receptor agonist with high potency and selectivity. *Journal of Pharmacology Experimental Therapeutics, 270,* 1270–1281.

Foreman, M. M., & Hall, J. L. (1987). Effects of D2 dopaminergic receptor stimulation on the lordotic response of female rats. *Psychopharmacology (Berlin), 91,* 96–100.

Fukuda, H., & Fukai, K (1986a). Ascending and descending pathways of reflex straining in the dog. *Japanese Journal of Physiology, 36,* 905–920.

Fukuda, H., & Fukai, K. (1986b). Location of the reflex centre for straining elicited by activation of pelvic afferent fibres of decerebrate dogs. *Brain Research, 380,* 287–296.

Fukuda, H., & Fukai, K. (1988). Discharges of bulbar respiratory neurons during rhythmic straining evoked by activation of pelvic afferent fibers in dogs. *Brain Research, 449,* 157–166.

Fukuda, H., Fukai, K., Yamane, M., & Okada, H. (1981). Pontine reticular unit responses to pelvic nerve and colonic mechanical stimulation in the dog. *Brain Research, 207,* 59–71.

Gerhardt, C. C., & van Heerikhuizen, H. (1997). Functional characteristics of heterologously expressed serotonin receptors. *European Journal of Pharmacology, 334,* 1–23.

Giuliano, F., Allard, J., Rampin, O., Droupy, S., Benoit, G., Alexandre, L., & Bernabe, J. (2002). Pro-erectile effect of systemic apomorphine: Existence of a spinal site of action. *Journal of Urology, 167,* 402–406.

Giuliano, F., Bernabe, J., McKenna, K., Longueville, F., & Rampin, O. (2001). Spinal proerectile effect of oxytocin in anesthetized rats. *American Journal of Physiology, 280,* R1870–R1877.

Giuliano, F., Rampin, O., Brown, K., Courtois, F., Benoit, G., & Jardin, A. (1996). Stimulation of the medial preoptic area of the hypothalamus in the rat elicits increases in intracavernous pressure. *Neuroscience Letters, 209,* 1–4.

Goldstein, A., & Hansteen, R. W. (1977). Evidence against involvement of endorphine in sexual arousal and orgasm in man. *Archives of General Psychiatry, 34,* 1179–1180.

Goldstein, I. (2000). Oral phentolamine: An alpha-1, alpha-2 adrenergic antagonist for the treatment of erectile dysfunction. *International Journal of Impotence Research, 12*(Suppl. 1), S75–S80.

Goldstein, I., Carson, C., Rosen, R., & Islam, A. (2001). Vasomax for the treatment of male erectile dysfunction. *World Journal of Urology, 19,* 51–56.

Goldstein, J. A. (1986). Erectile function and naltrexone. *Annals of Internal Medicine, 105,* 99.

Goodwin, F. K. (1971). Psychiatric side effects of levodopa in man. *Journal of the American Medical Association, 218*, 1915–1920.

Gréco, B., Edwards, D. A., Michael, R. P., & Clancy, A. N. (1996). Androgen receptor immunoreactivity and mating-induced Fos expression in forebrain and midbrain structures in the male rat. *Neuroscience, 75*, 161–171.

Gwinup, G. (1988). Oral phentolamine in non-specific erectile insufficiency. *Annals of Internal Medicine, 109*, 162–163.

Hamburger-Bar, R., & Rigter, H. (1975). Apomorphine: Facilitation of sexual behaviour in female rats. *European Journal of Pharmacology, 32*, 357–360.

Hancock, M. B., & Peveto, C. A. (1979a). A preganglionic autonomic nucleus in the dorsal gray commissure of the lumbar spinal cord of the rat. *Journal of Comparative Neurology, 183*, 65–72.

Hancock, M. B., & Peveto, C. A. (1979b). Preganglionic neurons in the sacral spinal cord of the rat: An HRP study. *Neuroscience Letters, 11*, 1–5.

Hart, B. L., & Leedy, M. G. (1985). Neurological bases of male sexual behaviour: A comparative analysis. In N. Adler, D. Pfaff, & R. W. Goy (Eds.), *Handbook of behavioral neurobiology: Vol. 7. Reproduction* (pp. 373–422). New York: Plenum Press.

Hasson, H. M. (1993). Cervical removal at hysterectomy for benign disease: Risks and benefits. *Journal of Reproductive Medicine, 38*, 781–790.

Hayes, E. S., Doherty, P. C., Hanson, L. A., Gorzalka, B. B., & Adaikan, P. G. (2000). Proerectile effects of novel serotonin agonists. *International Journal of Impotence Research, 12*(Suppl. 3), S62.

Heaton, J. P. (2001). Apomorphine: An update of clinical trial results. *International Journal of Impotence Research, 12*(Suppl. 4), S67–S73.

Hernandez-Gonzalez, M., Oropeza, M. V., Guevara, M. A., Cervantes, M., & Morali, G. (1994). Effects of intrathecal administration of adrenergic agonists on the frequency of copulatory pelvic thrusting of the male rat. *Archives of Medical Research, 25*, 419–425.

Holstege, G., Griffiths, D., De Wall, H., & Dalm, E. (1986). Anatomical and physiological observations on supraspinal control of bladder and urethral sphincter muscles in the cat. *Journal of Comparative Neurology, 250*, 449–461.

Holstege, G., Kuypers, H. G. J. M., & Boer, R. C. (1969). Anatomical evidence for direct brain stem projections to the somatic motoneuronal cell groups and autonomic preganglionic cell groups in cat spinal cord. *Brain Research, 171*, 329–333.

Holstege, J. C. (1987). Brainstem projections to lumbar motoneurons in rat: II. An ultrastructural study by means of the anterograde transport of wheat germ agglutinin coupled to horseradish peroxidase and using the tetramethyl benzidine reaction. *Neuroscience, 21*, 369–376.

Honda, C. N. (1985). Visceral and somatic afferent convergence onto neurons near the central canal in the sacral spinal cord of the cat. *Journal of Neurophysiology, 11*, 1059–1076.

Hubscher, C. H., & Johnson, R. D. (1996). Responses of medullary reticular formation neurons to input from the male genitalia. *Journal of Neurophysiology, 76*, 2474–2482.

Hughes, A. M., Everitt B. J., & Herbert, J. (1987). Selective effects of beta-endorphin infused into the hypothalamus, preoptic area and bed nucleus of the stria terminalis on the sexual and ingestive behavior of male rats. *Neuroscience, 23*, 1063–1073.

Hull, E. M. (1995). Dopaminergic influences on male rat sexual behavior. In P. Micevych & R. P. Hammer (Eds.), *Neurobiological effects of sex steroid hormones* (pp. 234–253). Cambridge, England: Cambridge University Press.

Hull, E. M., Bitran, D., Pehek, E. A., Warner, R. K., Band, L. C., & Holmes, G. M. (1986). Dopaminergic control of male sex behavior in rats: Effects of an intracerebrally-infused agonist. *Brain Research, 370*, 73–81.

Hull, E. M., Du, J., Lorrain, D. S., & Matuszewich, L. (1995). Extracellular dopamine in the medial preoptic area: Implications for sexual motivation and hormonal control of copulation. *Journal of Neuroscience, 15*, 7465–7471.

Hull, E. M., Eaton, R. C., Markowski, V. P., Moses, J., Lumley, L. A., & Loucks, J. A. (1992). Opposite influence of medial preoptic D1 and D2 receptors on genital reflexes: Implications for copulation. *Life Science, 51*, 1705–1713.

Jänig, W., & McLachlan, E. M. (1987). Organization of lumbar spinal outflow to distal colon and pelvic organs. *Physiological Reviews, 67*, 1332–1403.

Jordan, C. L., Breedlove, S. M., & Arnold, A. P. (1982). Sexual dimorphism and the influence of neonatal androgen in the dorsolateral motor nucleus of the rat lumbar spinal cord. *Brain Research, 249*, 309–314.

Julien, E., & Over, R. (1984). Male sexual arousal with repeated exposure to erotic stimuli. *Archives of Sexual Behavior, 13*, 211–221.

Kluver, H., & Bucy, P. C. (1939). Preliminary analysis of the functions of the temporal lobes in monkeys. *Archives of Neurology and Psychiatry, 42*, 979–1000.

Kojima, M., Matsuura, T., Tanaka, A., Amagai, T., Imanishi, J., & Sano, Y. (1985). Characteristic distribution of noradrenergic terminals on the anterior horn motoneurons innervating the perineal striated muscles in the rat. *Anatomy and Embryology, 171*, 267–273.

Kollar, E. J. (1953). Reproduction in the female rat after pelvic neurectomy. *Anatomical Record, 115*, 641–658.

Komisaruk, B. R., Adler, N. T., & Hutchinson, J. (1972). Genital sensory field: enlargement by estrogen treatment in female rats. *Science, 178*, 1295–1298.

Komisaruk, B. R., Gerdes, C. A., & Whipple, B. (1997). "Complete" spinal cord injury does not block perceptual responses to genital self-stimulation in women. *Archives of Neurology, 54*, 1513–1520.

Kondo, Y., Yamanouchi, K., & Arai, Y. (1993). p-Chlorophenylalanine facilitates copulatory behavior in septal lesioned but not in preoptic lesioned male rats. *Journal of Neuroendocrinology, 5*, 629–633.

Kow, L.-M., & Pfaff, D. W. (1973). Effects of estrogen treatment on the size of receptive field and response threshold of pudendal nerve in the female rat. *Neuroendocrinology, 13,* 299–313.

Kuru, M. (1965). Nervous control of micturition *Physiology Review, 45,* 425.

Kuypers, H. G. J. M., & Ugolini, G. (1990). Viruses as transneuronal tracers. *Trends in Neurosciences, 13,* 71–75.

Lal, S., Ackman, D., Thavundayil, J. X., Kiely, M. E., & Etienne, P. (1984). Effect of apomorphine, a dopamine receptor agonist, on penile tumescence in normal subjects. *Progress in Neuropsychopharmacology and Biological Psychiatry, 8,* 695–699.

Lal, S., Laryea, E., Thavundayil, J. X., Nair, N. P., Negrete, J., Ackman, D., et al. (1987). Apomorphine-induced penile tumescence in impotent patients: Preliminary findings. *Progress in Neuropsychopharmacology and Biological Psychiatry, 11,* 235–242.

Lane, R. M. (1997). A critical review of selective serotonin reuptake inhibitor-related sexual dysfunction; incidence, possible aetiology and implications for management. *Journal of Psychopharmacology, 11,* 72–82.

Langworthy, O. R. (1965). Innervation of the pelvic organs of the rat. *Investigative Urology, 2,* 491–511.

Lee, J. W., & Erskine, M. S. (1996). Vaginocervical stimulation suppresses the expression of c-fos induced by mating in thoracic, lumbar and sacral segments of the female rat. *Neuroscience, 74,* 237–249.

Lindstrom, S., Fall, M., Carlsson, C. A., & Erlandson, B. E. (1983). The neurophysiological basis of bladder inhibition in response to intravaginal electrical stimulation. *Journal of Urology, 29,* 405–410.

Lindvall, O., Björklund, A., & Skagerberg, G. (1984). Selective histochemical demonstration of dopamine terminal systems in rat di- and telencephalon: New evidence for dopaminergic innervation of hypothalamic neurosecretory nuclei. *Brain Research, 306,* 19–30.

Loewy, A. D., & McKellar, S. (1981). Serotonergic projections from the ventral medulla to the intermediolateral cell column in the rat. *Brain Research, 211,* 146–152.

Loewy, A. D., McKellar, S., & Saper, C. B. (1979). Direct projections from the A5 catecholamine cell group to the intermediolateral cell column. *Brain Research, 174,* 309–314.

Mains, R. E., Eippers, B. A., & Ling, N. (1977). Common precursor to the corticotropins and endorphins. *Proceedings of the National Academy of Sciences USA, 74,* 3014–3018.

Malsbury, C. W. (1971). Facilitation of male rat copulatory behavior by electrical stimulation of the medial preoptic area. *Physiology and Behavior, 7,* 797–805.

Marlier, L., Teilhac, J. R., Cerruti, C., & Privat, A. (1991). Autoradiographic mapping of serotonin 1, serotonin 1A, serotonin 1B and serotonin 2 receptors in the rat spinal cord. *Brain Research, 550,* 15–23.

Marson, L. (1995). Central nervous system neurons identified after injection of pseudorabies virus into the rat clitoris *Neuroscience Letters, 190,* 41–44.

Marson, L., List, M. S., & McKenna, K. E. (1992). Lesions of the nucleus paragigantocellularis alter ex copula penile reflexes. *Brain Research, 592,* 187–192.

Marson, L., & McKenna, K. E. (1990). The identification of a brainstem site controlling spinal sexual reflexes in male rats. *Brain Research, 515,* 303–308.

Marson, L., & McKenna, K. E. (1992). A role for 5-hydroxytryptamine in descending inhibition of spinal sexual reflexes. *Experimental Brain Research, 88,* 313–320.

Marson, L., & McKenna, K. E. (1994). Stimulation of the hypothalamus initiates the urethrogenital reflex in male rats. *Brain Research, 638,* 103–108.

Marson, L., & McKenna, K. E. (1996). CNS cell groups involved in the control of the ischiocavernosus and bulbospongiosus muscles: A transneuronal tracing study using pseudorabies virus. *Journal of Comparative Neurology, 374,* 161–179.

Marson, L., Platt, K. B., & McKenna, K. E. (1993). CNS innervation of the penis as revealed by the transneuronal transport of pseudorabies virus. *Neuroscience, 55,* 263–280.

Martin, G. F., Vertes, R. P., & Waltzer, R. (1985). Spinal projections of the gigantocellular reticular formation in the rat: Evidence for projections from different areas to laminae I and II and lamina IX. *Experimental Brain Research, 58,* 154–162.

Matuszewich, L., Lorrain, D. S., & Hull, E. M. (2000). Dopamine release in the medial preoptic area of female rats in response to hormonal manipulation and sexual activity. *Behavioral Neuroscience, 114,* 772–782.

Matuszewich, L., Lorrain, D. S., Trujillo, R., Dominguez, J., Putnam, S. K., & Hull, E. M. (1999). Partial antagonism of 8-OH-DPAT's effects on male rat sexual behavior with a D2, but not a 5-HT1A, antagonist. *Brain Research, 820,* 55–62.

McIntosh, T. K., & Barfield, R. J. (1984). Brain monoaminergic control of male reproductive behavior: I. Serotonin and the postejaculatory period. *Behavioral Brain Research, 12,* 255–265.

McIntosh, T. K., Vallano, M. L., & Barfield, R. J. (1980). Effects of morphine, beta-endorphin and naloxone on catecholamine levels and sexual behavior in the male rat. *Pharmacology Biochemistry and Behavior, 13,* 435–441.

McKenna, K. E., Chung, S. K., & McVary, K. T. (1991). A model for the study of sexual function in anesthetized male and female rats. *American Journal of Physiology, 261,* R1276–R1285.

McKenna, K. E., & Nadelhaft, I. (1986). The organization of the pudendal nerve in the male and female rat. *Journal of Comparative Neurology, 248,* 532–549.

McKenna, K. E., & Nadelhaft, I. (1989). The pudendo-pudendal reflex in male and female rats. *Journal of Autonomic Nervous System, 27,* 67–77.

Meisel, R. L., & Sachs, B. D. (1994). The physiology of male sexual behavior. In E. Knobil & J. D. Neill (Eds.), *The physiology of reproduction* (pp. 3–105). New York: Raven Press.

Melis, M. R., & Argiolas, A. (1995). Nitric oxide donors induce penile erection and yawning when injected in the central nervous system of male rats. *European Journal of Pharmacology, 294,* 1–9.

Melis, M. R., Argiolas, A., & Gessa, G. L. (1987). Apomorphine-induced penile erection and yawning: Site of action in the brain. *Brain Research, 415,* 98–104.

Melis, M. R., Argiolas, A., & Gessa, G. L. (1989). Evidence that apomorphine induces penile erection and yawning by releasing oxytocin in the central nervous system. *European Journal of Pharmacology, 164,* 565–570.

Melis, M. R., Stancampiano, R., Gessa, G. L., & Argiolas, A. (1992). Prevention by morphine of apomorphine- and oxytocin-induced penile erection: Site of action in the brain. *Neuropsychopharmacology, 6,* 17–21.

Melis, M. R., Succu, S., & Argiolas, A. (1997). Prevention by morphine of N-methyl-D-aspartic acid-induced penile erection and yawning: Involvement of nitric oxide. *Brain Research Bulletin, 44,* 689–694.

Melis, M. R., Succu, S., Iannucci, U., & Argiolas, A. (1997a). Oxytocin increases nitric oxide production in the paraventricular nucleus of the hypothalamus of male rats: Correlation with penile erection and yawning. *Regulatory Peptides, 69,* 105–111.

Melis, M. R., Succu, S., Iannucci, U., & Argiolas, A. (1997b). Prevention by morphine of apomorphine- and oxytocin-induced penile erection and yawning: Involvement of nitric oxide. *Naunyn Schmiedebergs Archives of Pharmacology, 355,* 595–600.

Melis, M. R., Succu, S., Iannucci, U., & Argiolas, A. (1998). Nitric oxide production is increased in the paraventricular nucleus of the hypothalamus of male rats during non-contact penile erections and copulation. *European Journal of Neuroscience, 10,* 1968–1974.

Merari, A., & Ginton, A. (1975). Characteristics of exaggerated sexual behavior induced by electrical stimulation of the medial preoptic area in male rats. *Brain Research, 86,* 97–108.

Millan, M. J., Peglion, J. L., Lavielle, G., & Perrin-Monneyron, S. (1997). Serotonin 2C receptors mediate penile erections in rats: Actions of novel and selective agonists and antagonists. *European Journal of Pharmacology, 325,* 9–12.

Miller, N. S., & Gold, M. S. (1988). The human sexual response and alcohol and drugs. *Journal Substance Abuse Treatment, 5,* 171–177.

Modell, J. G., Katholi, C. R., Modell, J. D., & DePalma, R. L. (1997). Comparative sexual side effects of bupropion, fluoxetine, paroxetine, and sertraline. *Clinical Pharmacology and Therapeutics, 61,* 476–487.

Monsma, F. J., Shen, Y., & Ward, R. P. (1993). Cloning and expression of a novel serotonin receptor with high affinity for tricyclic psychotropic drugs. *Molecular Pharmacology, 43,* 320–327.

Morales, A. (2000). Yohimbine in erectile dysfunction: The facts. *International Journal of Impotence Research, 12*(Suppl. 1), S70–S74.

Morales, A., Condra, M., & Owen, J. A. (1987). Is yohimbine effective in the treatment of organic impotence? Results of a controlled trial. *Journal of Urology, 137,* 1168–1172.

Murphy, M. R., Seckl, J. R., Burton, S., Checkley, S. A., & Lightman, S. L. (1987). Changes in oxytocin and vasopressin secretion during sexual activity in men. *Journal of Clinical Endocrinology and Metabolism, 65,* 738–741.

Nadelhaft, I., & Booth, A. M. (1984). The location and morphology of preganglionic neurons and the distribution of visceral afferents from the rat pelvic nerve: A horseradish peroxidase study. *Journal of Comparative Neurology, 226,* 238–245.

Nadelhaft, I., & McKenna, K. E. (1987). Sexual dimorphism in sympathetic preganglionic neurons of the rat hypogastric nerve. *Journal of Comparative Neurology, 256,* 308–315.

Nygren, L. G., & Olson, L. (1977). A new major projection from locus ceruleus: The main source of noradrenergic nerve terminals in the ventral and dorsal columns of the spinal cord. *Brain Research, 132,* 85–93.

Papka, R. E., McCurdy, J. R., Williams, S. J., Mayer, B., Marson, L., & Platt, K. B. (1995). Parasympathetic preganglionic neurons in the spinal cord involved in uterine innervation are cholinergic and nitric oxide-containing. *Anatomical Record, 241,* 554–562.

Papka, R. E., Williams, S., Miller, K. E., Copelin, T., & Puri, P. (1998). CNS location of uterine-related neurons revealed by trans-synaptic tracing with pseudorabies virus and their relation to estrogen receptor-immunoreactive neurons. *Neuroscience, 84,* 935–952.

Park, K., Goldstein, I., Andry, C., Siroky, M. B., Krane, R. J., & Azadzoi, K. M. (1997). Vasculogenic female sexual dysfunction: The hemodynamic basis for vaginal engorgement insufficiency and clitoral erectile insufficiency. *International Journal of Impotence Research, 9,* 27–37.

Pehek, E. A., Thompson, J. T., & Hull, E. M. (1989a). The effects of intrathecal administration of the dopamine agonist apomorphine on penile reflexes and copulation in the male rat. *Psychopharmacology (Berlin), 99,* 304–308.

Pehek, E. A., Thompson, J. T., & Hull, E. M. (1989b). The effects of intracranial administration of the dopamine agonist apomorphine on penile reflexes and seminal emission in the rat. *Brain Research, 500,* 325–332.

Pescatori, E. S., Engelman, J. C., Davis, G., & Goldstein, I. (1993). Priapism of the clitoris: A case-report following tradazone use. *Journal of Urology, 149,* 1557–1559.

Pfaus, J. G., & Gorzalka, B. B. (1987). Opioids and sexual behaviour. *Neuroscience and Biobehavior Reviews, 11,* 1–34.

Poggioli, R., Arletti, R., Benelli, A., Cavazzuti, E., & Bertolini, A. (1998). Diabetic rats are unresponsive to the penile erection-inducing effect of intracerebroventricularly injected adrenocorticotropin. *Neuropeptides, 32,* 151–155.

Pomerantz, S. M., Hepner, B. C., & Wertz, J. M. (1993). Serotonergic influences on male sexual behavior of rhesus monkeys: Effects of serotonin agonists. *Psychopharmacology (Berlin)*, *111*, 47–54.

Purinton, P. T., Fletcher, T. F., & Bradley, W. E. (1976). Innervation of pelvic viscera in the rat. *Investigative Urology*, *14*, 28–32.

Rampin, O., Bernabe, J., & Giuliano, F. (1997). Spinal control of penile erection. *World Journal of Urology*, *15*, 2–13.

Rattner, W. H., Gerlaugh, R. L., Murphy, J. J., & Erdman, W. J., II. (1958). The bulbocavernosus reflex: I. Electromyographic study of normal patients. *Journal of Urology*, *80*, 140–141.

Reid, K., Morales, A., Harris, C., Surridge, D. H., Condra, M., & Owen, J. (1987). Double blind trial of yohimbine in treatment of psychogenic impotence. *Lancet*, *2*(8556), 421–423.

Ridet, J. L., Tamir, H., & Privat, A. (1994). Direct immunocytochemical localization of 5-hydroxytryptamine receptors in the adult rat spinal cord: A light and electron microscopic study using an anti-idiotypic antiserum. *Journal of Neuroscience Research*, *38*, 109–121.

Rodriguez-Manzo, G., & Fernandez-Guasti, A. (1994). Reversal of sexual exhaustion by serotonergic and noradrenergic agents. *Behavioral Brain Research*, *62*, 127–34.

Rose, J. D. (1990). Brainstem influences on sexual behavior. In W. R. Klemm & R. P. Vertes (Eds.), *Brainstem mechanisms of behavior* (pp. 407–463). New York: Wiley.

Saenz de Tejada, I., Ware, J. C., Blanco, R., Pittard, J. T., Nadig, P. W., Azadzoi, K. M., et al. (1991). Pathophysiology of prolonged penile erection associated with trazodone use. *Journal of Urology*, *145*, 60–64.

Sagar, S. M., Sharp, F. R., & Curran, T. (1988). Expression of c-fos protein in brain: Metabolic mapping at the cellular level. *Science*, *240*, 1326–32.

Saito, S., Kidd, G. J., Trapp, B. D., Dawson, T. M., Bredt, D. S., Wilson, D. A., et al. (1994). Rat spinal cord neurons contain nitric oxide synthase. *Neuroscience*, *59*, 447–456.

Saper, C. B., Loewy, A. D., Swanson, L. W., & Cowan, W. M. (1976). Direct hypothalamo-autonomic connections. *Brain Research*, *117*, 305–312.

Sar, M., & Stumpf, W. E. (1975). Distribution of androgen concentrating neurons in the rat brain. In W. E. Stumpf & L. D. Grant (Eds.), *Anatomical neuroendocrinology* (pp. 120–133). Basel, Switzerland: Karger.

Sato, Y., Christ, G. J., Horita, H., Adachi, H., Suzuki, N., & Tsukamoto, T. (1999). The effects of alterations in nitric oxide levels in the paraventricular nucleus on copulatory behavior and reflexive erections in male rats. *Journal of Urology*, *162*, 2182–2185.

Sato, Y., Horita, H., Kurohata, T., Adachi, H., & Tsukamoto, T. (1998). Effect of the nitric oxide level in the medial preoptic area on male copulatory behavior in rats. *American Journal of Physiology*, *274*, R243–R247.

Sawchenko, P. E., & Swanson, L. W. (1982). Immunohistochemical identification of neurons in the paraventricular nucleus of the hypothalamus that project to the medulla or to the spinal cord in the rat. *Journal of Comparative Neurology, 205*, 260–272.

Schrøder, H. D. (1980). Organization of the motoneurons innervating the pelvic muscles of the male rat. *Journal of Comparative Neurology, 192*, 567–587.

Schrøder, H. D., & Skagerberg, G. (1985). Catecholamine innervation of the caudal spinal cord in the rat. *Journal of Comparative Neurology, 242*, 358–368.

Segraves, R. T., Bari, M., Segraves, K., & Spirna, P. (1991). Effect of apomorphine on penile tumescence in men with psychogenic impotence. *Journal of Urology, 145*, 1174–1175.

Shimura, T., & Shimokochi, M. (1991). Modification of male rat copulatory behavior by lateral midbrain stimulation. *Physiology and Behavior, 50*, 989–994.

Simerly, R. B., & Swanson, L. W. (1986). The organization of neural inputs to the medial preoptic nucleus of the rat. *Journal of Comparative Neurology, 246*, 312–342.

Simerly, R. B., & Swanson, L. W. (1988). Projections of the medial preoptic nucleus: Phaseolus vulgaris leucoagglutinin anterograde tract-tracing study in the rat. *Journal of Comparative Neurology, 270*, 209–242.

Sipski, M. L., Alexander, C. J., & Rosen, R. C. (1985). Orgasm in women with spinal cord injuries: A laboratory-based assessment. *Archives of Physical Medicine and Rehabilitation, 76*, 1097–1102.

Skagerberg, G., & Bjorklund, A. (1985). Topographic principles in the spinal projections of serotonergic and non-serotonergic brainstem neurons in the rat. *Neuroscience, 15*, 445–480.

Skagerberg, G., & Lindvall, O. (1985). Organization of diencephalic dopamine neurons projecting to the spinal cord of the rat. *Brain Research, 342*, 340–341.

Skagerberg, G., Bjorklund, A., Lindvall, O., & Schmidt, R. H. (1982). Origin and termination of the diencephalo-spinal dopamine system in the rat. *Brain Research Bulletin, 9*, 237–244.

Slimp, J. C., Hart, B. L., & Goy, R. W. (1975). Heterosexual, autosexual and social behavior of adult male rhesus monkeys with medial preoptic-anterior hypothalamic lesions. *Brain Research, 142*, 105–122.

Steers, W. D., & de Groat, W. C. (1989). Effects of m-clorophenylpiperazine on penile and bladder function in rats. *American Journal of Physiology, 257*, R1441–R1449.

Steers, W. D., Mallory, B., & de Groat, W. C. (1988). Electrophysiological study of neural activity in penile nerve of the rat. *American Journal of Physiology, 254*, R989–R1000.

Steinbusch, H. (1981). Distribution of serotonin-immunoreactivity in the central nervous system of the rat-cell bodies and terminals. *Neuroscience, 6*, 557–618.

Svensson, L., & Hansen, S. (1984). Spinal monoaminergic modulation of masculine copulatory behavior in the rat. *Brain Research, 302*, 315–321.

Swanson, L. W., & Kuypers, H. G. J. M. (1980). The paraventricular nucleus of the hypothalamus: Cytoarchitectonic subdivisions and organization of projections to the pituitary, dorsal vagal complex, and spinal cord as demonstrated by retrograde fluorescence double-labeling methods. *Journal of Comparative Neurology, 194,* 555–570.

Szele, F. G., Murphy, D. L., & Garrick, N. A. (1988). Effects of fenfluramine, mchlorophenylpiperazine, and other serotonin-related agonists and antagonists on penile erections in nonhuman primates. *Life Science, 43,* 1297–1303.

Tang, Y., Rampin, O., Calas, A., Facchinetti, P., & Giuliano, F. (1998). Oxytocinergic and serotonergic innervation of identified lumbosacral nuclei controlling penile erection in the male rat. *Neuroscience, 82,* 241–254.

Tang, Y., Rampin, O., Giuliano, F., & Ugolini, G. (1999). Spinal and brain circuits to motoneurons of the bulbospongiosus muscle: Retrograde transneuronal tracing with rabies virus. *Journal of Comparative Neurology, 414,* 167–192.

Tetel, M. J., Getzinger, M. J., & Blaustein, J. D. (1993). Fos expression in the rat brain following vaginal-cervical stimulation by mating and manual probing. *Journal of Neuroendocrinology, 5,* 397–404.

Theodosis, D. T. (1985). Oxytocin-immunoreactive terminals synapse on oxytocinergic neurons in the supraoptic nuclei. *Nature, 313,* 682–684.

Thor, K. B., Nickolaus, S., & Helke, C. J. (1993). Autoradiographic localization of 5-hydroxytryptamine1A, 5-hydroxytryptamine1B and 5-hydroxytryptamine1C/2 binding sites in the rat spinal cord. *Neuroscience, 55,* 235–252.

Truitt, W. A., & Coolen, L. M. (2002). Identification of a potential ejaculation generator in the spinal cord. *Science, 297,* 1566–1569.

Uitti, R. J., Tanner, C. M., Rajput, A. H., Goetz, C. G., Klawans, H. L., & Thiessen, B. (1989). Hypersexuality with antiparkinsonian therapy. *Clinical Neuro-Pharmacology, 12,* 375–383.

Valtschanoff, J. G., Weinberg, R. J., & Rustioni, A. (1992). NADPH diaphorase in the spinal cord of rats. *Journal of Comparative Neurology, 321,* 209–222.

Van Bockstaele, E. J., Pieribone, V. A., & Aston-Jones, G. (1989). Diverse afferents converge on the nucleus paragigantocellularis in the rat ventrolateral medulla: Retrograde and anterograde tracing studies. *Journal of Comparative Neurology, 290,* 561–584.

Van der Ploeg, L. H., Martin, W. J., Howard, A. D., Nargund, R. P., Austin, C. P., Guan, X., et al. (2002). A role for the melanocortin 4 receptor in sexual function. *Proceedings of the National Academy of Sciences USA, 99,* 11381–11386.

Van Dis, H., & Larsson, K. (1970). Seminal discharge following intracranial electrical stimulation. *Brain Research, 23,* 381–386.

Veening, J. G., & Coolen, L. M. (1998). Neural activation following sexual behavior in the male and female rat brain. *Behavioral Brain Research, 92,* 181–93.

Véronneau-Longueville, F., Rampin, O., Freund-Mercier, M.-J., Tang, Y., Calas, A., Marson, L., et al. (1999). Oxytocinergic innervation of autonomic nuclei controlling penile erection in the rat. *Neuroscience, 93,* 1437–1447.

Vodusek, D. B. (1990). Pudendal SEP and bulbocavernosus reflex in women. *Electroencephalography and Clinical Neurophysiology, 77*, 134–136.

Wagner, C. K., & Clemens, L. G. (1991). Projections of the paraventricular nucleus of the hypothalamus to the sexually dimorphic lumbosacral region of the spinal cord. *Brain Research, 539*, 254–262.

Wagner, C. K., & Clemens, L. G. (1993). Neurophysin-containing pathway from the paraventricular nucleus of the hypothalamus to a sexually dimorphic motor nucleus in lumbar spinal cord. *Journal of Comparative Neurology, 336*, 106–116.

Waldinger, M. D. (2002). The neurobiological approach to premature ejaculation. *Journal of Urology, 168*, 2359–2367.

Waldinger, M. D., Berendsen, H. H. G., Blok, B. F., Olivier, B., & Holstege, G. (1998). Premature ejaculation and serotonergic antidepressants-induced delayed ejaculation: The involvement of the serotonergic system. *Behavioral Brain Research, 92*, 111–118.

Walsh, P. C., & Donker, P. J. (2002). Impotence following radical prostatectomy: Insight into etiology and prevention. *Journal of Urology, 167*, 1005–1010.

Warner, R. K., Thompson, J. T., Markowski, V. P., Loucks, J. A., Bazzett, T. J., Eaton, R. C., & Hull, E. M. (1991). Microinjection of the dopamine antagonist cis-flupenthixol into the MPOA impairs copulation, penile reflexes and sexual motivation in male rats. *Brain Research, 540*, 177–182.

Wessells, H., Hruby, V. J., Hackett, J., Han, A. G., Balse-Srinivasan, P., & Vanderah, T. W. (2003). Ac-Nle-c[Asp-His-DPhe-Arg-Trp-Lys]-NH2 induces penile erection via brain and spinal melanocortin receptors. *Neuroscience, 118*, 755–762.

Wessels, H., Fuciarelli, K., Hansen, J., Hadley, M. E., Hruby, V. J., Dorr, R., & Levine, N. (1998). Synthetic melanotropic peptide initiates erections in men with psychogenic erectile dysfunction: Double blind, placebo controlled crossover study. *Journal of Urology, 160*, 389–393.

Wessels, H., Levine, N., Hadley, M. E., Dorr, R., & Hruby, V. (2000). Melanocortin receptor agonists, penile erection, and sexual motivation: Human studies with Melanotan II. *International Journal of Impotence Research, 12*(Suppl 4), S74–S79.

Whipple, B., Richards, E., Tepper, M., & Komisaruk, B. R. (1995). Sexual responses in women with complete spinal cord injury. In D. M. Krotosku, M. Nosek, & M. Turk (Eds.), *The health of women with physical disabilities: Setting a research agenda for the 90's*. Baltimore: Paul H. Brookes.

Wikberg, J. E. (1999). Melanocortin receptors: Perspectives for novel drugs. *European Journal of Pharmacology, 375*, 295–310.

Wikberg, J. E., Muceniece, R., Mandrika, I., Prusis, P., Lindblom, J., Post, C., & Skottner, A. (2000). New aspects on the melanocortins and their receptors. *Pharmacological Research, 42*, 393–420.

Yanagimoto, M., Honda, K., Goto, Y., & Negoro, H. (1996). Afferents originating from the dorsal penile nerve excite oxytocin cells in the hypothalamic paraventricular nucleus of the rat. *Brain Research, 733*, 292–296.

Yang, S. P., & Voogt, J. L. (2001). Mating-activated brainstem catecholaminergic neurons in the female rat. *Brain Research, 894,* 159–166.

Yells, D. P., Hendricks, S. E., & Prendergast, M. A. (1992). Lesions of the nucleus paragigantocellularis: Effects on mating behavior in male rats. *Brain Research, 596,* 73–79.

Zorgniotti, A. W. (1992). "On demand" erection with oral preparations for impotence: 3- (N- (2-imidazoline-2ylmethyl) -p-toluidinol) phenol mesylate. *International Journal of Impotence Research, 4*(Suppl 2), A99.

Zorgniotti, A. W. (1994). Experience with buccal phentolamine mesylate for impotence. *International Journal of Impotence Research, 6,* 37–41.

Zorgniotti, A. W., & Lizza, E. F. (1994). Effect of large doses of the nitric oxide precursor, L-arginine, on erectile dysfunction. *International Journal of Impotence Research, 6,* 33–35.

5

BRAIN ACTIVITY IMAGING DURING SEXUAL RESPONSE IN WOMEN WITH SPINAL CORD INJURY

BARRY R. KOMISARUK AND BEVERLY WHIPPLE

The present series of brain-imaging studies was initiated in response to intriguing early reports that women diagnosed with "complete" spinal cord injury may experience orgasms during sexual intercourse (Cole, 1975; Kettl et al., 1991; Whipple, 1990), more recent laboratory confirmatory reports (Komisaruk & Whipple, 1994; Sipski & Alexander, 1995; Sipski, Alexander, & Rosen, 1995; Whipple, Gerdes, & Komisaruk, 1996), and a report that pregnant women with spinal cord injury below T12 can feel uterine contractions and movement of their fetus in utero (Berard, 1989). The reports are surprising because the genital sensory pathways that ascend from the genitalia to the brain are presumably interrupted by the spinal cord injury. According to the diagnosis of complete spinal cord injury,

It is with the deepest gratitude that we acknowledge our expert collaborators on the brain imaging project: for the positron emission tomography studies, John W. Keyes, Beth Harkness, and Carolyn Gerdes; and for the functional magnetic resonance imaging studies, Kris Mosier, Andrew Kalnin, Wen-Ching Liu, Audrita Crawford, and Sherry Grimes. Funding support was from The Christopher Reeve Paralysis Foundation, National Institutes of Health (Grant R25GM60826), and The Charles and Johanna Busch Foundation, Rutgers, The State University of New Jersey.

women do not have sensibility below the level of the injury. Indeed, as some of the women in the studies that we describe in this chapter reported to us, their physicians told them that their "sensations" are imaginary, a reply that they said upset them.[1]

On the basis of descriptions of the peripheral distribution and level of entry into the spinal cord of the genital sensory nerves in women (Bonica, 1967), and our and others' mapping of the sensory fields and zones of entry into the spinal cord of the genital sensory nerves in the female rat (Berkley, Hotta, Robbins, & Sato, 1990; Cunningham, Steinman, Whipple, Mayer, & Komisaruk, 1991; Komisaruk, Adler, & Hutchison, 1972; Kow & Pfaff, 1973–1974; Peters, Kristal, & Komisaruk, 1987; see also McKenna, chap. 4, this volume), we suspected that the location along the course of the spinal cord at which the injury occurred is critical. We hypothesized that some genital sensation could occur even if the complete spinal cord injury extended as high as spinal cord thoracic level 11. We formulated this hypothesis on the basis that the hypogastric nerves ascend in the sympathetic chain and enter the spinal cord at thoracic levels 10–12 (Bonica, 1967; Felten & Jozefowicz, 2003).

BACKGROUND

To provide context and rationale for the brain imaging studies described below, we first describe the classic picture of the different nerves that innervate the specific genital regions.

The division of labor among the genital sensory nerves is basically as follows. The hypogastric nerves convey sensory activity from the uterus and cervix in women (Bonica, 1967) and in rats (Berkley et al., 1990; Peters et al., 1987); the pelvic nerves convey sensory activity from the cervix, vagina, and midline perineal skin (Berkley et al., 1990; Komisaruk et al., 1972; Peters et al., 1987); and the pudendal nerves convey sensory activity from the clitoris and perineal skin (Peters et al., 1987). Thus, hypogastric nerve fibers that enter the spinal cord at T10, directly above a complete injury as high as T11, could conceivably convey some genital sensory activity that could access the brain and hence attain perceptibility. The pudendal and pelvic nerves enter the spinal cord at upper sacral and lower lumbar levels (Ding et al., 1999).

---

[1] The clinical diagnosis of "complete" spinal cord injury is based on criteria by the American Spinal Injury Association (1992) of absence of sensation of pinprick (pain) or cotton wisp (touch) and absence of voluntary movement, all below the level of the injury, plus the absence of sensation of rectal digital stimulation.

Laboratory Study of Women With Complete Spinal Cord Injury at Specific Levels

From the above considerations, it is reasonable to infer that cervical stimulation, accessing the brain via activation of the hypogastric nerves, might be sufficient to elicit orgasm. To test this hypothesis and assess the division of labor among the genital sensory nerves, we analyzed perceptual responses to genital self-stimulation in women who were diagnosed with complete spinal cord injury that was between the level of entry of the hypogastric and pelvic-pudendal nerves (thereby allowing access to the brain of the former but not the latter). We compared this group with able-bodied women and women diagnosed with complete spinal cord injury at or above T10. We hypothesized that this latter group would show no responses to vaginal or cervical self-stimulation. However, to our surprise, rather than showing no responses, the women with complete spinal cord injury at or above T10 showed the strongest responses to the vaginal or cervical self-stimulation, compared with the other two groups of women. Our unexpected findings led us to hypothesize that genital sensory responses in these women can be conveyed via the Vagus nerves (Komisaruk & Whipple, 1994; Komisaruk, Gerdes, & Whipple, 1997; Whipple et al., 1996). We have tested this hypothesis by using functional brain imaging methodology—positron emission tomography (PET) and functional magnetic resonance imaging (fMRI)—to ascertain whether cervical self-stimulation would activate the sensory projection nucleus of the Vagus nerve in women with complete injury to, or transection of, the spinal cord (Komisaruk, Whipple, et al., 2002; Whipple & Komisaruk, 2002).

The Vagus Nerves

The findings that led us to hypothesize a genital sensory role for the Vagus nerves were as follows. We used several measures to ascertain responses to vaginal or cervical self-stimulation. In addition to assessing the amount of force at which the women reported they could feel the stimulator pressing against the cervix (Komisaruk, Gerdes, & Whipple, 1997), whether they experienced sexual responses or orgasm from the stimulation (Whipple et al., 1996), and, independently of the laboratory setting, whether they perceived menstrual discomfort, we also obtained an objective measure—pain thresholds—measured at the fingers.[2]

---

[2] Why measure pain thresholds? Our previous studies in the laboratory rat showed that vaginocervical pressure produces a pain blockage so effective that the rats do not withdraw from the painful stimulus (e.g., an intensely hot lamp) beyond the point at which they would sustain tissue damage (Komisaruk et al., 1996). During natural mating, the female rat's pain threshold increases to a level more than that produced by three times the analgesic dose of morphine (Gomora, Beyer, Gonzalez-

We measured pain thresholds before, during, and after cervical self-stimulation by applying a calibrated, gradually increasing force via a 1-mm-tip diameter plastic rod to the fingers up to a maximum force of 1 kg. We recorded the force at which the participant reported that it felt painful, which we took as the *pain-detection threshold*. We also measured tactile thresholds using a calibrated, graded series of von Frey fibers applied to the dorsal surface of the hand (Komisaruk, Gerdes, & Whipple, 1997; Whipple & Komisaruk, 1985, 1988).

Rationale for the Research Methodology Used

The basis for using these pain and touch threshold determinations was as follows. We had shown previously (Komisaruk & Whipple, 1984; Whipple & Komisaruk, 1985) that when noninjured women applied a steady force against the anterior vaginal wall ("Grafenberg spot"), their pain detection thresholds increased by an average of more than 50% over control, resting baseline levels. At the same time, the tactile thresholds showed no significant change. When the women in that study applied vaginal self-stimulation in a manner that they reported felt pleasurable, their average pain detection threshold increased more than 75%, with no significant change in tactile thresholds. Four of the 10 women in the study experienced orgasm during the vaginal self-stimulation, and under that condition, the average pain detection threshold increased by more than 100% over the resting-control baseline levels, and again, there was no significant change in tactile thresholds (Whipple & Komisaruk, 1985). A typical subjective report by the women was that they were aware of the increasing compressive force on their fingers, but it did not bother them until it actually became painful (but at that point the average compressive force averaged at least 50% greater than before the vaginal self-stimulation). Thus, the women were fully aware of the testing situation at all times, suggesting that the elevation in pain thresholds was not due to distraction. Another line of evidence that they were not distracted from the test situation by the vaginal self-stimulation is that they provided reports on their (unchanging)

Mariscal, & Komisaruk, 1994), and during parturition—the other event when vaginocervical pressure stimulation occurs normally—pain thresholds are also increased (Toniolo, Whipple, & Komisaruk, 1987).

We speculate that the function of this pain blockage during (a) parturition is to attenuate the stress of parturition and facilitate bonding between mother and young and (b) mating, to render the female willing to accept the multiple intromissions that are necessary to stimulate the secretion of levels of progesterone that are necessary to prepare the uterus for implantation of the fertilized ova. These findings led us to test and demonstrate that a comparable phenomenon of pain blockage in response to vaginal or cervical stimulation exists in women (for review, see Komisaruk & Whipple, 1995).

tactile sensitivity continually throughout the testing period, including the time when they were applying the vaginal self-stimulation (Whipple & Komisaruk, 1985).

In a second study, we used additional methods to assess the possibility that distraction could account for the findings; that is, the participants watched the train-chase scene from the film, *The French Connection*, or they brushed a fur mitt in a pleasurable manner against their skin, while pain thresholds and tactile thresholds were measured. Although a significant increase in pain detection thresholds during vaginal self-stimulation was again observed in this study, there was no significant effect of any of the distracting stimuli on pain detection thresholds and no change in tactile thresholds (Whipple & Komisaruk, 1985). From these findings we concluded that vaginal self-stimulation produces analgesia, that is, a selective attenuation of pain but not touch. After completion of the day's testing, when each woman was asked what she thought we were testing, almost all answered that we were trying to ascertain whether pain interferes with sexual response. Because this is the opposite of what we were actually testing, it is also unlikely that their preconceptions led to bias in an expectation of analgesia.

Indirect Evidence Supporting a "Genitosensory Vagus" Hypothesis

Returning to consideration of our study on responses to cervical self-stimulation in women with complete spinal cord injury (Komisaruk, Gerdes, & Whipple, 1997; Whipple et al., 1996), the control group in that study consisted of 5 women without spinal cord injury. The experimental groups consisted of 10 women with complete spinal cord injury below T10 and 6 women with complete spinal cord injury above the level of entry into the spinal cord of all three pairs of these nerves (i.e., complete injury at T10 or above). We expected that the women in the latter group (i.e., complete injury at T10 or above) would show no response to the self-stimulation. We measured pain detection thresholds and tactile thresholds as in our previous studies in women. Our unexpected and surprising finding was that, contrary to our expectation that the women with complete spinal cord injury at T10 or above would fail to respond to the self-stimulation, the women with this highest level of spinal cord injury actually had the strongest analgesic responses to vaginal or cervical self-stimulation among the three groups. All 6 of the women also stated that they commonly experienced menstrual discomfort, which originated from below the level of their spinal cord injury. One of the 6 women in this group reported several orgasms in the laboratory during the cervical self-stimulation. Four of the 6 women were able to feel the stimulator when it was pressed against the cervix, although the force threshold for their feeling it was significantly higher than that in the noninjured women (Komisaruk, Gerdes, & Whipple, 1997; Komisaruk & Whipple, 1994; Whipple et al., 1996).

Evidence for a "Genitosensory Vagus" Based on Studies in Laboratory Animals

To account for this finding—and on the basis of a report in rats by Guevara-Guzman and colleagues (Ortega-Villalobos et al., 1990) that the neural tracer, horseradish peroxidase, when injected into the cervix produced labeling of neurons in the nodose ganglion, which is the dorsal root (i.e., sensory) ganglion of the Vagus nerve—we hypothesized that in the women with complete spinal cord injury, the basis for their perceptual responses to the cervical self-stimulation was that the Vagus nerves were providing the sensory pathway for the cervical stimulation to reach the brain. The innervation of the uterus and cervix by the Vagus nerves in the rat was subsequently confirmed by Papka and colleagues (Collins, Lin, Berthoud, & Papka, 1999).

Vagal electrical stimulation has been shown to produce analgesia in rats (Maixner & Randich, 1984; Ness, Randich, Fillingim, Faught, & Backensto, 2001; Randich & Gebhart, 1992) and in humans (Kirchner, Birklein, Stefan, & Handwerker, 2000), and we reported that vaginocervical probing in rats produces analgesia even after bilateral transection of the known genitospinal nerves (pudendal, pelvic, and hypogastric). That analgesia is abolished after subsequent bilateral transection of the Vagus nerves (Bianca et al., 1994; Cueva-Rolon et al., 1994, 1996).

We tested the Vagus nerve hypothesis in rats using two different approaches: transecting the known genitospinal nerves or transecting the spinal cord, followed by testing for persistence of any responses to vaginocervical stimulation (VCS), followed by bilaterally transecting the Vagus nerves and subsequent retesting for responses to VCS. VCS produced analgesia after bilateral transection of the pudendal, pelvic, and hypogastric nerves, although the magnitude of the analgesia was lower than in the intact rats; subsequent bilateral transection of the Vagus nerves in the abdomen abolished this analgesia (Cueva-Rolon et al., 1994, 1996).

In a separate study, we found that significant pupil dilatation in response to VCS persisted, although at a diminished magnitude, after total surgical ablation of the spinal cord at the mid-thoracic level (T7) of the spinal cord in rats. Subsequent bilateral transection of the Vagus nerves in the abdomen abolished the pupil dilatation response to VCS (Komisaruk, Bianca, et al., 1996).

As a further test of this Vagus nerve hypothesis, we transected one of the pair of Vagus nerves in the neck and stimulated a central end electrically. We observed an immediate and marked pupil dilatation, thus supporting our hypothesis that vagal afferent activity stimulates pupil dilatation (Komisaruk et al., 1996).[3]

Hypothesis

Our hypothesis is that vaginocervical self-stimulation in women with total interruption of the spinal cord will activate the brain region to which the Vagus nerve directly projects. The sensory fibers of the Vagus nerves pass totally outside the spinal cord, through the abdominal cavity, the diaphragm, the thoracic cavity, and the neck. The primary afferent terminals of these fibers project directly to the nucleus of the solitary tract (NTS), which is situated in the medulla oblongata of the lower brainstem. Thus, the NTS is the site of the first synapse between the sensory fibers of the Vagus nerve and the central nervous system. Hubscher and Berkley (1994, 1995) reported that neurons of the NTS in rats responded to mechanical stimulation of the vagina, cervix, uterus, or rectum, and vagotomy altered these responses. On the basis of the above studies, we concluded that the Vagus nerves constitute a functional afferent pathway from the vagina and cervix that bypasses the spinal cord, projecting directly to the brain, specifically to the NTS of the lower brainstem (Komisaruk, Gerdes, & Whipple, 1997; Komisaruk & Whipple, 1994, 2000; Whipple et al., 1996).

NEW RESEARCH FINDINGS: TESTING THE HYPOTHESIS

To test whether the NTS is activated by vaginocervical self-stimulation in women with the highest level of complete spinal cord injury and in noninjured women, we have used positron emission tomography (PET) with O^{15}-labeled water as the tracer (Komisaruk, Whipple, Gerdes, Harkness, & Keyes, 1997) and functional magnetic resonance imaging (fMRI; Komisaruk, Whipple, et al., 1996, 1997, 2002; Whipple & Komisaruk, 2002). The resolution of the PET method appears to be too low to localize responses in the medulla. As an alternative, we have found that the fMRI method has higher resolution plus adequate sensitivity to show brain responses to

---

[3] There is a possible source of confusion regarding the Vagus nerve as a parasympathetic nerve and the fact that parasympathetic stimulation constricts, rather than dilates, the pupil. There is, however, no contradiction with the present findings, for neither of these considerations is relevant to the present findings. That is, first, the parasympathetic innervation of the eye is via the oculomotor nerve (cranial nerve III) rather than the Vagus. And second, it is afferent, not efferent, activity of the Vagus nerve that is producing the pupil dilatation. Thus, Vagus nerve afferent activity, via stimulation of its central nervous system projection pathways, has the capacity to stimulate a sympathetic-dominated response of the pupil.

genital self-stimulation. Using the fMRI method, we have obtained evidence that cervical self-stimulation produces activation of NTS in women with complete spinal cord injury above T10. These findings support our hypothesis that the Vagus nerve provides a spinal cord bypass pathway in women for afferent activity generated by cervical self-stimulation.

The PET Method: Findings, Advantages, and Limitations

Despite the relatively low resolving power of the PET method, we have nevertheless observed activation in what appears to be the region of the paraventricular nucleus (PVN) of the hypothalamus in a noninjured woman during orgasm that she induced by cervical self-stimulation. Our subsequent studies using fMRI have confirmed this response in additional women. These findings are consistent with reports of release of oxytocin into plasma during orgasm in women (Blaicher et al., 1999; Carmichael et al., 1987; Carmichael, Warburton, Dixen, & Davidson, 1994).

Figures 5.1 through 5.3 show schematically our hypothesis of the location of the neural pathways that mediate vaginal or cervical self-stimulation-induced analgesia, measured at the fingers, in women with intact or transected spinal cord. In Figure 5.1, the radial nerve conveys noxious sensory input to the spinal cord at the cervical level, where it synapses on second-order neurons of the spinothalamic system and then on third-order neurons that project from thalamus to cortex.

Figure 5.2 is based on the likelihood of similarity with the endogenous analgesia-producing system in rats (reviewed in Komisaruk & Whipple, 2000). This illustration depicts our concept of a system by which genitospinal afferents via the pelvic and hypogastric nerves synapse first in the spinal cord and then project to the lower brainstem, where they stimulate descending noradrenergic (NE) and serotoninergic (5-HT) pathways that generate presynaptic inhibition of the primary afferents of the radial nerve. This would result in inhibiting the release of pain transmitters and blocking the nociceptive activity.

Figure 5.3 shows the proposed vagal afferent pathway that bypasses the spinal cord, terminates in the NTS in the medulla oblongata, which is the vagal sensory nucleus, and stimulates the same descending pathway to block the nociceptive input at the primary afferent terminals of the radial nerve in the cervical level of the spinal cord.

We hypothesized that in the women with complete interruption of the spinal cord, cervical self-stimulation produces analgesia by activating the Vagus nerves, and consequently that the NTS would be activated by this stimulation. We first used PET with O^{15}-labeled water as the tracer and superimposed the PET activity images on anatomical MRI to enable neuroanatomical localization of the activity. The principle of action of this

Finger pain pathway

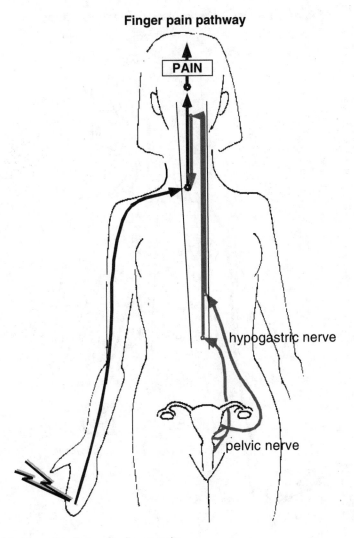

Figure 5.1. Neural pathway for finger pain.

method is that the more active a cluster of neurons becomes, the greater is the amount of blood that becomes supplied to that cluster, and thus, the greater is the amount of radioactivity per unit time in the region of the cluster. A greater concentration of radioactivity in a brain region appears as a brighter region in the images (these images use a "hot-metal" gradient representation, in which dark red is "coolest," representing lowest radioactivity, and yellow-white is "hottest," representing highest radioactivity).

Figure 5.4 shows the activity in the region of the NTS during cervical self-stimulation in a woman with complete spinal cord injury at T8. There

Vaginal-cervical self-stimulation attenuates finger pain

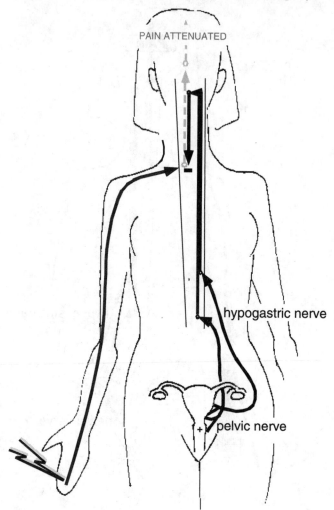

Figure 5.2. Presumptive genitospinal neural pathways mediating pain attenuation. NE = noradrenergic pathway; 5-HT = serotoninergic pathway.

are two control conditions: the first, resting with no cervical self-stimulation, and the second, application of heated vibrators to both feet simultaneously. Note that in the region delineated by the oval, there is no visible activity during the resting control condition. Furthermore, in the foot stimulation control condition, there is little or no visible activity. The rationale for this control condition is that the modalities of touch, pressure, vibration, and heat are all stimulated concurrently, and absence of activation of the thalamus would confirm the completeness of spinal cord injury. Analysis of the

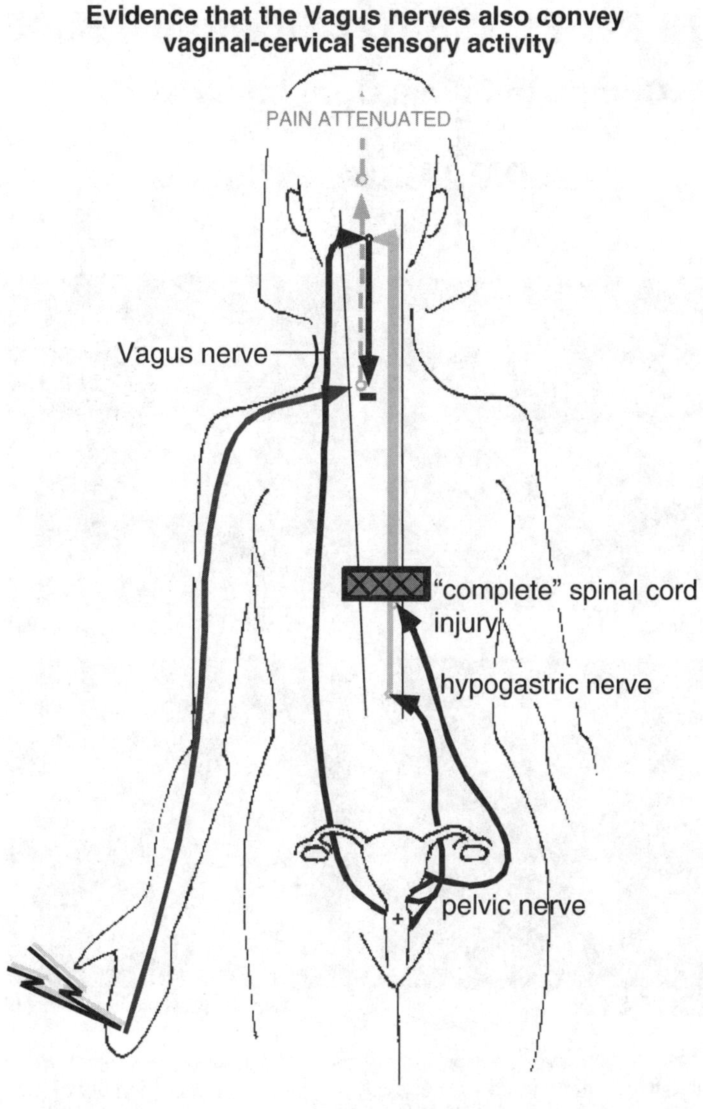
Evidence that the Vagus nerves also convey
vaginal-cervical sensory activity
PAIN ATTENUATED
Vagus nerve
"complete" spinal cord
injury
hypogastric nerve
pelvic nerve

Figure 5.3. Proposed spinal-cord-bypass pathway through which pain is attenuated by vaginocervical stimulation.

thalamic activity did in fact show little or no activation, by contrast with a noninjured woman subjected to the same stimulation. By contrast with these two control conditions, there is substantial activity in the delineated region (NTS) during cervical self-stimulation. A similar response was obtained in a second woman with complete spinal cord injury at T9 (Komisaruk, Whipple, et al., 1996; Whipple & Komisaruk, 2002). A major limitation of the PET method is its relatively low resolving power. That is, we

NTS RESPONSE TO CERVICAL SELF-STIM.

Complete Spinal Cord Injury at T8

COMBINED MRI - PET SCANS

Subject # 233 / 1298
Transaxial view

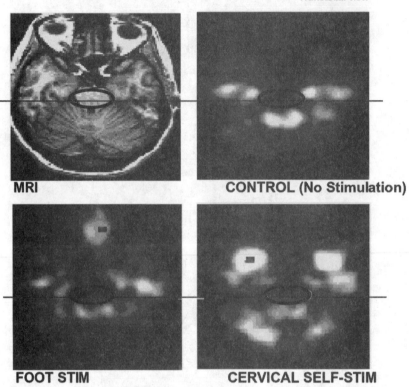

Figure 5.4. Positron emission tomography (PET) imaging of brainstem response to cervical self-stimulation in a woman with complete spinal cord injury at T8. The white ovals encircle the same region of the brain in each image. The upper left image is the anatomical image of the brain; the other three are the PET images. Note that the highest activity occurred during cervical self-stimulation. NTS = nucleus of the solitary tract; STIM = stimulation; MRI = magnetic resonance imaging.

can see that the general region of the NTS in the medulla is active, but we cannot localize the region of activity restricted to the NTS.

Nevertheless, the resolving power of the PET method was adequate to enable visualization of heightened activity in certain larger brain regions in the case of a noninjured woman who experienced an orgasm during the scan. In this case we observed heightened activity in the midbrain reticular

MIDBRAIN RETICULAR FORMATION

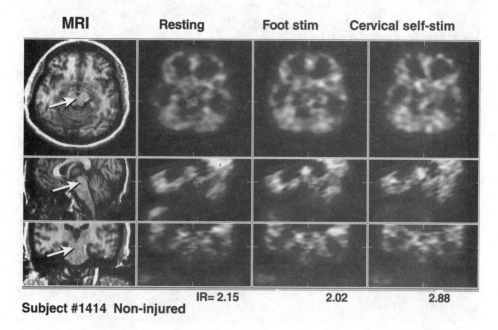

| MRI | Resting | Foot stim | Cervical self-stim |

Subject #1414 Non-injured

IR= 2.15   2.02   2.88

Figure 5.5. Positron emission tomography (PET) imaging of midbrain reticular formation during orgasm and control conditions in an uninjured woman. The circles are located at the same region in each of the images. The arrows in the anatomical images in the leftmost column point to the same region. Note that the highest activity occurred during cervical self-stimulation. The top row is a transaxial view, the middle row a sagittal view, and the bottom row a coronal view, all of the same brain region. MRI = magnetic resonance imaging; stim = stimulation; IR = the ratio of the quantified image intensity in the region of interest (delineated by a circle) to that in an adjacent control brain region (white matter) in each same image. The same conventions are used here as in Figures 5.6 and 5.7, following.

formation (which others, using PET, have found to show increased activity during arousal; Balkin et al., 2002), basal ganglia (particularly the region of the putamen), and anterior hypothalamus in the region of the PVN (Figures 5.5, 5.6, and 5.7). The format of these figures shows the MRI anatomical brain "slice" (from top to bottom at the leftmost portion of the figures) in transaxial (i.e., horizontal), sagittal, and coronal orientation, respectively. The corresponding activity patterns are shown during the control resting period, foot stimulation, and cervical self-stimulation, respectively, from the left to the right side of each figure. The intensity of activity is depicted using the "hot metal" representation. The numerical values shown at the bottom of each set are the ratios between the activity in the "region of interest" shown as a circle in each slice and the activity in a

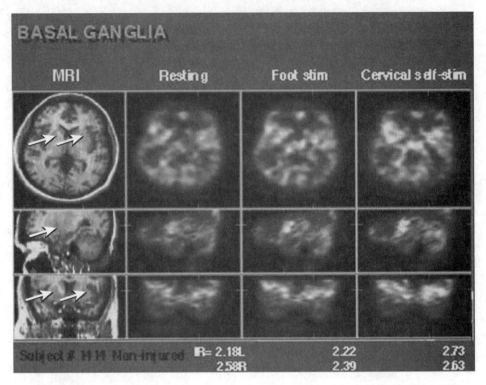

Figure 5.6. Positron emission tomography (PET) imaging of the basal ganglia during orgasm and control conditions in an uninjured woman. MRI = magnetic resonance imaging; stim = stimulation; IR = the ratio of the quantified image intensity in the region of interest to that in an adjacent control brain region in the same image.

control, equivalent-sized region of white matter in the same "slice." All the circles in each figure—MRI and PET—are in registration with each other; that is, they all circumscribe the same brain region, regardless of the orientation of the slices. Note that the ratios are highest during orgasm.

It is perhaps not surprising that the midbrain reticular formation and the basal ganglia are most active during orgasm, for they are involved in general arousal and in gross motor control, respectively. However, it was unexpected, but is particularly interesting, that the region of the hypothalamic PVN shows high activity. Oxytocin is synthesized in neurons of the PVN and is transported through the axons of these neurons, to the axon terminals in the posterior pituitary gland (i.e., the neurohypophysis; for a review, see, e.g., Cross & Wakerley, 1977). There, the oxytocin is released by conventional action potential activity from the axon terminals of these "neuroendocrine neurons" into the systemic circulation. Several studies have reported that oxytocin levels in the systemic circulation rise significantly during orgasm in women and men (Blaicher et al., 1999; Carmichael et al.,

PARAVENTRICULAR NUCLEUS of HYPOTHALAMUS

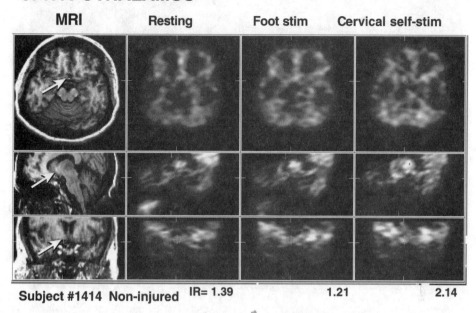

| MRI | Resting | Foot stim | Cervical self-stim |

Subject #1414 Non-injured   IR= 1.39   1.21   2.14

Figure 5.7. Positron emission tomography (PET) imaging of the hypothalamus during orgasm and control conditions in an uninjured woman. MRI = magnetic resonance imaging; stim = stimulation; IR = the ratio of the quantified image intensity in the region of interest to that in an adjacent control brain region in the same image.

1987, 1994). We have reported in rats that VCS immediately releases oxytocin into the systemic circulation (and into the spinal cord; Sansone et al., 2002). It has been suggested that one of the functions of the oxytocin in the systemic circulation is to stimulate contractions of the uterus, which facilitate the transport of seminal fluid through an inward (i.e., toward the uterus) suction effect (Wildt, Kissler, Licht, & Becker, 1998). These data are the first of which we are aware providing direct evidence that the PVN is activated during orgasm.

fMRI: A Higher Resolution Brain Imaging Method

To ascertain whether the NTS is activated by cervical self-stimulation in women who have complete spinal cord injury above the level of entry into the cord of the genitospinal nerves (i.e., the pudendal, pelvic, and hypogastric nerves), we needed a functional brain imaging method that has higher resolving power than the PET method. We first established the greater resolving power of the fMRI method in the lower brainstem by

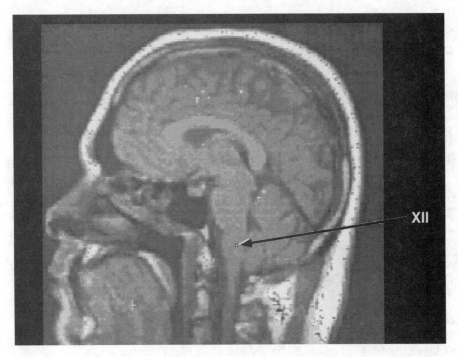

Figure 5.8. Functional magnetic resonance imaging (fMRI) can resolve activity in localized brain regions: Activation of the region of the hypoglossal nucleus by tongue movement.

mapping cranial nerve nuclei in the vicinity of the NTS. For example, the hypoglossal nucleus, which is also in the medulla oblongata, controls tongue movement. Tapping the tongue against the roof of the mouth produced activation in the precise location of the hypoglossal nucleus (based on histological atlases; Figure 5.8), convincing us that the resolving power of the fMRI method is far better than that of the PET method, and that it is adequate to localize the NTS (Komisaruk, Mosier, et al., 2002).

fMRI Evidence

To aid in ascertaining the location of the NTS on the fMRI images, we used existing evidence that the upper (i.e., superior) level of the NTS conveys taste sensory activity (Travers & Norgren, 1995) by having the research participants taste a sauce that combined sweet, salty, sour, and bitter qualities, by combining sugar, salt, lemon juice, and dry mustard, and mapped the region of activation, an example of which is shown in Figure 5.9. We then asked two women with complete spinal cord injury above T10 to perform cervical self-stimulation, after having tasted the sauce. Figure 5.10 shows an MRI of the spinal cord injury in the first woman. The

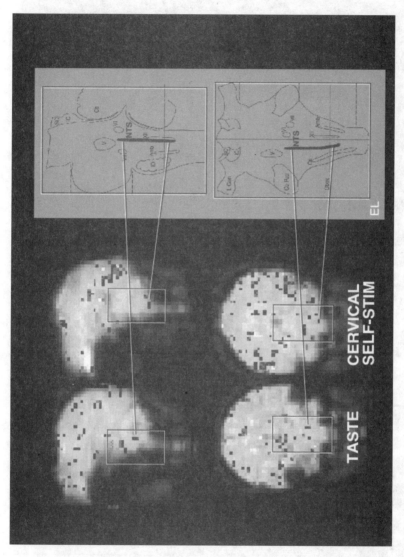

Figure 5.9. Functional magnetic resonance imaging of the sensory projection zone of the Vagus nerve—region of the nucleus of the solitary tract (NTS)—localized by a taste stimulus and activated by cervical self-stimulation (SELF-STIM). The activation by cervical self-stimulation supports our hypothesis that the Vagus nerve conveys cervical sensory activity to the brain, bypassing the spinal cord injury (SCI).

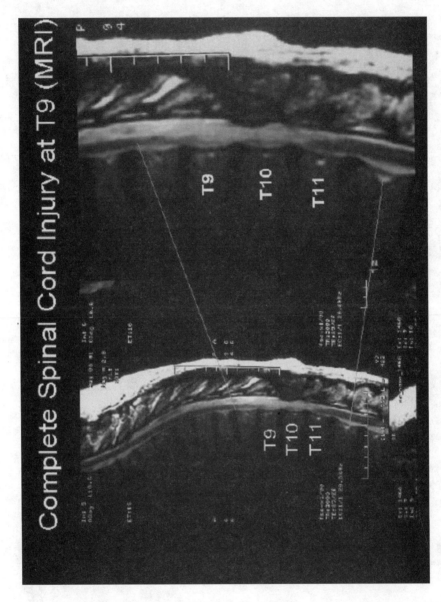

Figure 5.10. Magnetic resonance imaging (MRI) of the complete spinal cord injury of the woman in Figure 5.9.

fMRI of the woman (Figure 5.9) shows the anatomical location of the NTS based on histological atlases, and also shows in sagittal (top) and coronal (bottom) views the regions activated by taste and by cervical self-stimulation, respectively.

Figures 5.11 and 5.12 show, for the second woman, the comparable spinal cord injury site and brain regions activated. Note that in both cases, the region activated by taste was directly above (superior to) the region activated by cervical self-stimulation. This provides the first direct evidence that cervical self-stimulation activates the NTS, thus supporting our hypothesis that the Vagus nerves convey sensory activity from the cervix in women whose afferent pathways through the spinal cord are interrupted.

fMRI Data

The two women with complete spinal cord injury above T10 reported that they experienced orgasms during the cervical self-stimulation. There were marked changes in activity in widespread regions of the brain at the time of the orgasms. Figure 5.13 provides an overall view of the activity in the brain at the beginning of cervical self-stimulation compared, in Figure 5.14, with the activity of the same brain regions at orgasm. Note the much greater activation in the lower brainstem, frontal cortex, and cerebellum.

Figure 5.15 shows, at two different imaging criteria ($p < .05$, dots, and $p < .01$, which has a higher threshold for significance of response), that brain regions activated during orgasm included hypothalamus, amygdala, cingulate cortex, and insular cortex. The images at the left of this figure show the MRI anatomical image of the same brain, on which the activity is superimposed, and at the far left, an atlas image of the same brain region, based on histological specimens, which aids in the anatomical identification of the structures.

Figure 5.16 shows fMRI images at two different brain regions, the hypothalamus (upper and lower images on the right) and, rostral to that, the preoptic/bed nucleus of the stria terminalis region (upper and lower images on the left). The upper images show the fMRI activity at orgasm, and the lower images show the same activity superimposed on the corresponding brain anatomy. Note the activation in the region of the PVN of the hypothalamus, amygdala, cingulate cortex, insular cortex, and region of the nucleus accumbens.

Figure 5.17 shows fMRI activity in the region of the PVN of the hypothalamus during orgasm. The schematic view on the right shows Netter's diagram (Felten & Jozefowicz, 2003) of this region, locating the PVN to the left and slightly below the anterior commissure. The anatomical MRI image shows the comparable region, the crosshairs identifying the anterior commissure. The image to the right shows the fMRI activity at orgasm

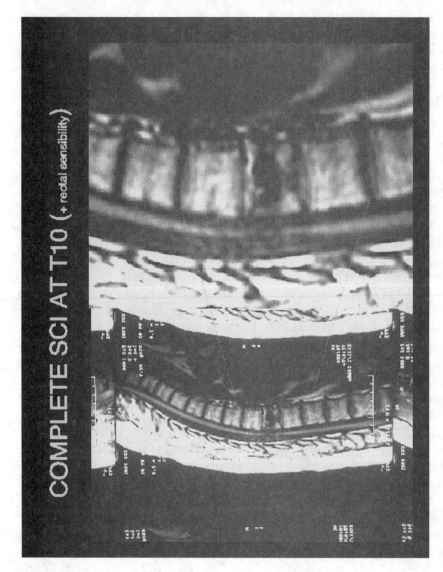

Figure 5.11. MRI of the complete spinal cord injury (SCI) in a second woman.

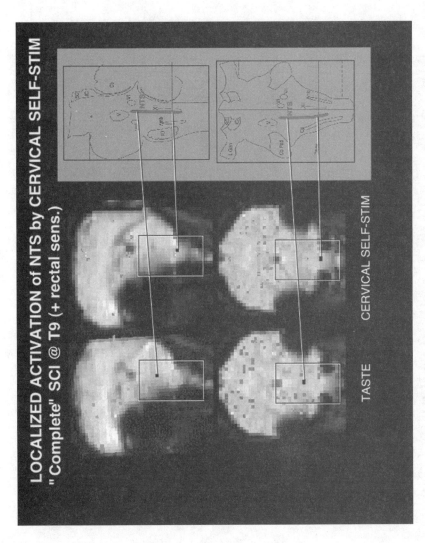

Figure 5.12. Functional magnetic resonance imaging of the woman in Figure 5.11, comparable with that in Figure 5.9. NTS = nucleus of the solitary tract; SELF-STIM = self-stimulation; SCI = spinal cord injury; sens. = sensation.

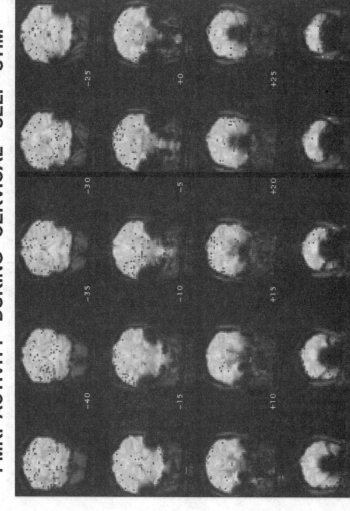

Figure 5.13. Functional magnetic resonance imaging (fMRI) of serial "sections" through the brain during cervical self-stimulation (SELF-STIM).

FMRI ACTIVITY DURING CERVICAL SELF -STIM DURING ORGASM

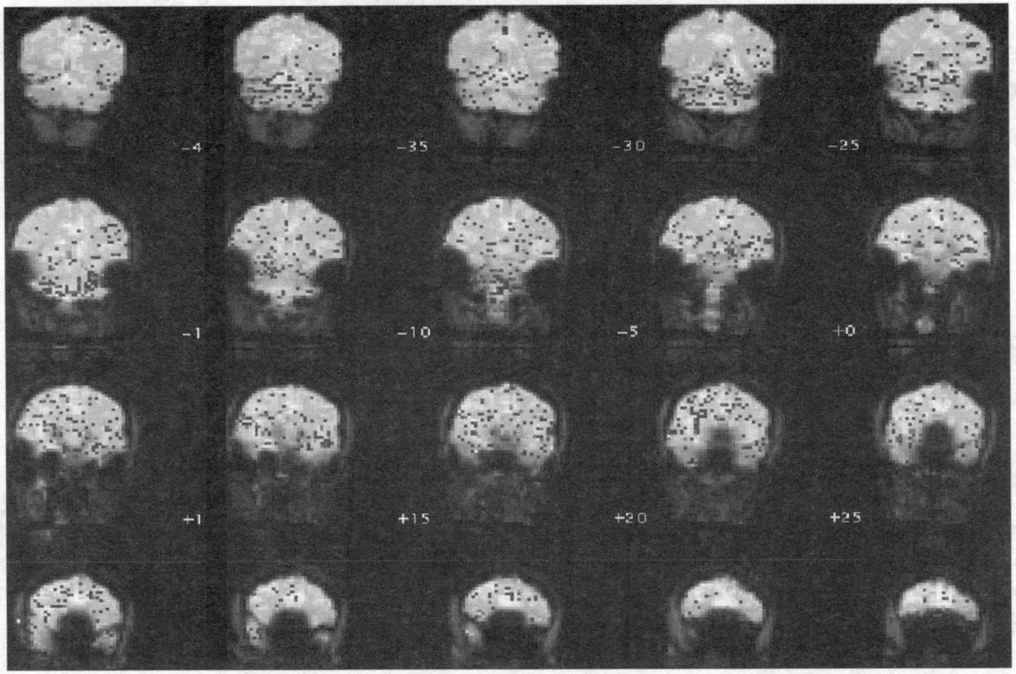

8901 Run2 #4

Figure 5.14. Functional magnetic resonance imaging (fMRI) of the same brain "sections" as in Figure 5.13, during orgasm induced by the cervical self-stimulation (SELF-STIM). Note the more widespread activation than in Figure 5.13, which was prior to orgasm.

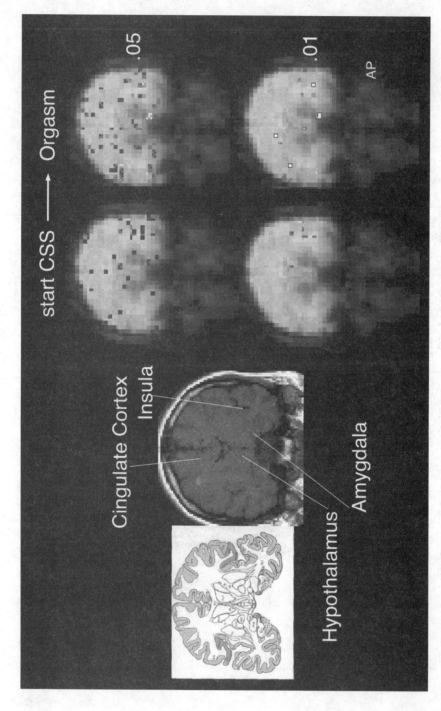

Figure 5.15. Functional magnetic resonance imaging showing specific regions of activation in the forebrain during orgasm. CSS = cervical self-stimulation.

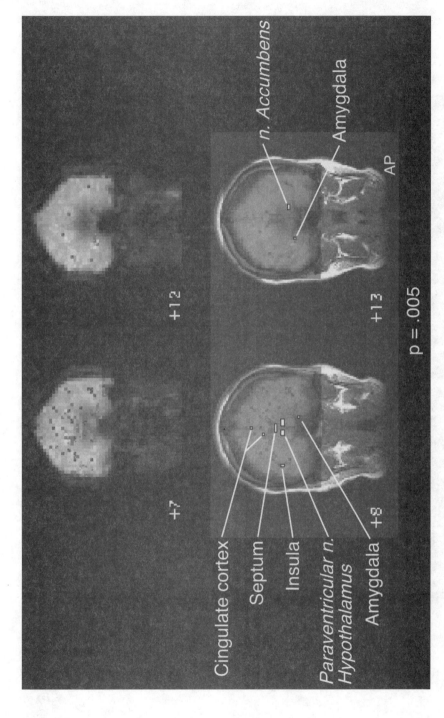

Figure 5.16. Functional magnetic resonance imaging showing activation of hypothalamus and adjacent regions during orgasm.

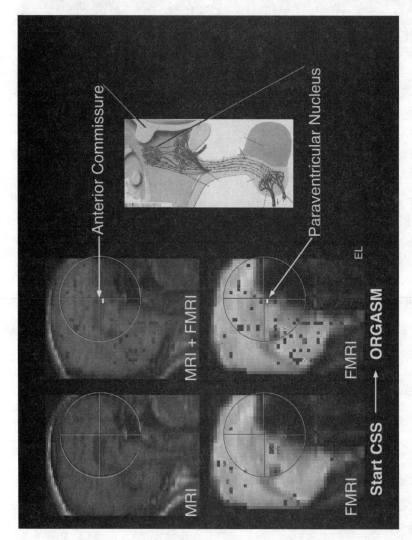

Figure 5.17. Functional magnetic resonance imaging (fMRI) imaging showing activation in the region of the paraventricular nucleus of the hypothalamus on orgasm. CSS = cervical self-stimulation. The artist's rendition of the region of the paraventricular nucleus is modified after Netter (Felten & Jozefowicz, 2003).

superimposed on this anatomical image. The two lower images show the fMRI only at the beginning of cervical self-stimulation (left) and at orgasm (right). Note the activation in the region of the PVN only at orgasm.

Figure 5.18 shows a greater activation of hippocampus at orgasm (pixels) than at the onset of cervical self-stimulation. The anatomical image shows the region of the hippocampus (oval), the superimposed fMRI activity at orgasm (image to its right), and the fMRI activity only before and during orgasm (lower and upper figures to the right, respectively).

DISCUSSION

The above findings, based on observation of brain activity, provide the first evidence to account for the intriguing reports that women with complete spinal cord injury can perceive genital sensory activity. The findings provide evidence of the mediating neural pathway and are, to our knowledge, also the first report of brain regions activated during orgasm in women.

The Vagus Nerves Convey Genital Sensory Activity in Women

The present findings of activation in the region of the NTS in women with complete spinal cord injury or total interruption of the spinal cord (verified anatomically using MRI) provide evidence that the NTS is activated by self-stimulation of the cervix (at the vaginal-uterine junction) in women who have complete spinal cord injury or total interruption of the spinal cord above the level of entry into the spinal cord of all the known genitospinal nerves. As in the case of the women in our previous study with complete spinal cord injury at comparable levels, these women reported that they experience menstrual discomfort, responded to cervical self-stimulation in a typical way by showing analgesia, and could perceive when the stimulator was pressed against the cervix, although their threshold of sensibility to this stimulus was significantly higher than that of women without spinal cord injury.

The Vagus nerves in women provide a genital sensory pathway that projects directly to the brain, bypassing the spinal cord, and consequently, spinal cord injury at any level. This pathway can, in some women with complete spinal cord injury, generate orgasm. We interpret these findings to support our hypothesis that the Vagus nerves, which project to the NTS, directly convey the afferent activity from the cervix, and thereby provide a sensory pathway from the cervix to the brain, bypassing the spinal cord.

Sexual response including orgasm has also been reported by Sipski et al. (1995) in women with complete spinal cord injury at the levels tested

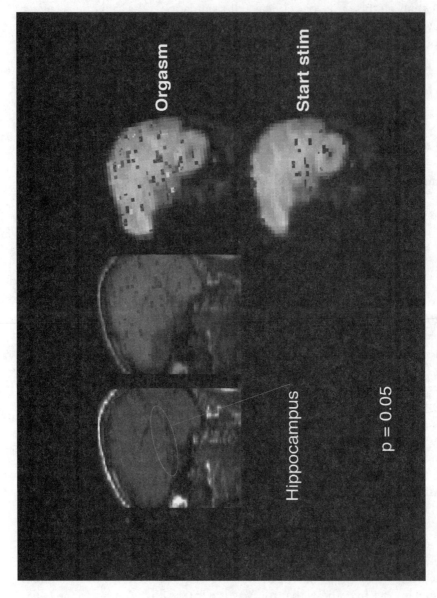

Figure 5.18. Functional magnetic resonance imaging showing activation of hippocampus on orgasm. stim = stimulation.

in the present study, and at higher spinal cord levels. Because the Vagus nerves enter the brain directly, this pathway could account for the findings in their study. Sipski et al. speculated that "autonomic pathways" mediate the effect, yet they provided no evidence to support that conclusion. More-over, they have not explicitly refuted our earlier evidence (Komisaruk & Whipple, 1994) that the Vagus nerves could provide an afferent pathway from the genitalia that would bypass the spinal cord at any level, including those that they report.

Multiple Brain Regions Are Activated During Orgasm in Women

The brain regions activated during orgasm that was elicited by cervical self-stimulation include hypothalamus, limbic system (including amygdala, hippocampus, cingulate cortex and insular cortex, and the region of the accumbens-bed nucleus of the stria terminalis-preoptic area), neocortex (including parietal and frontal cortices), basal ganglia (especially putamen), and cerebellum, in addition to lower brainstem (central gray, mesencephalic reticular formation, and NTS). Differences between regional activation during, versus before or after, orgasm suggest that areas more directly related to orgasm include paraventricular area of the hypothalamus, amygdala, anterior cingulate region of the limbic cortex, and region of the nucleus accumbens. There is no evidence of orgasm in female rats. However, some of the same brain regions have been reported to become activated during mating or VCS. Thus, using the c-fos immunocytochemical method, activation was reported in amygdala (Erskine & Hanrahan, 1997; Pfaus & Heeb, 1997; Rowe & Erskine, 1993; Tetel, Getzinger, & Blaustein, 1993; Veening & Coolen, 1998; Wersinger, Baum, & Erskine, 1993), PVN (Pfaus & Heeb, 1997; Rowe & Erskine, 1993), medial preoptic area (Erskine & Hanrahan, 1997; Tetel et al., 1993; Wersinger et al., 1993), midbrain central gray (Pfaus & Heeb, 1997; Tetel et al., 1993), and based on local release of dopamine, the nucleus accumbens (Pfaus, Damsma, Wenkstern, & Fibiger, 1995).

First Identification of Brain Regions Activated
During Orgasm in Women

To our knowledge, this is the first evidence of activation of hypothalamus during orgasm in men or women. Earlier reports of orgasm in men found activation in prefrontal cortex but not subcortical structures (Tiihonen et al., 1994). Also in men, Holstege and colleagues (Holstege et al., 2003) reported that the mesodiencephalic area, cerebellum pontine reticular formation, basal ganglia (putamen and claustrum), and several cortical regions including the lateral prefrontal cortex, but not the hypothalamus, were activated during orgasm. Furthermore, in men during sexual arousal, but not orgasm, elicited by photographs, Wallen and colleagues (Hamann, Herman, Nolan, & Wallen, 2002) reported that fMRI activity was increased in amygdala, hippocampus, and hypothalamus compared with the activity in women, whereas the striatal regions (caudate and nucleus accumbens) were activated in both men and women.

The findings in the present study of activation of PVN, observed using PET as well as fMRI, are consistent with reports of oxytocin release during orgasm. That is, the PVN neurons secrete oxytocin, which is stored in the posterior pituitary gland (Cross & Wakerley, 1977), vaginal or cervical stimulation releases the oxytocin from the posterior pituitary gland into the bloodstream—a response known as the Ferguson reflex (Ferguson, 1941)— and orgasm releases the oxytocin into the bloodstream (Blaicher et al., 1999; Carmichael et al., 1987, 1994; see also McKenna, chap. 4, this volume).

It is of interest that during orgasm, the insular cortex and anterior cingulate cortices are active, because both of these areas have been reported to be activated during response to pain (Bornhovd et al., 2002; Casey, Morrow, Lorenz, & Minoshima, 2001; Ploner, Gross, Timmermann, & Schnitzler, 2002). This suggests an interesting local interaction between the pain and pleasure regions of the brain. Further research is needed to compare directly and in the same individual brain regions that are activated during pleasure with those activated during pain. The nucleus accumbens also showed activation during orgasm in the present study. This brain region has been reported to show fMRI activation during the "rush" induced by an intravenous injection of nicotine (Stein et al., 1998). Our findings suggest a role for the nucleus accumbens in mediating orgasmic pleasure in women.

A salient and reliable feature of brain regions activated during orgasm was activation of the cerebellum. The cerebellum modulates muscle tension via the gamma efferent system and receives proprioceptive information (Felten & Jozefowicz, 2003). Because muscle tension can reach peak levels during orgasm (Masters & Johnson, 1966) and contribute to the sensory pleasure of orgasm (Komisaruk & Whipple, 1998, 2000), it is not unlikely that the cerebellum thereby plays a significant motoric and hedonic role in orgasm.

While genital self-stimulation is being applied continuously, certain localized brain regions become activated only when orgasm results from the stimulation (e.g., the PVN). There are at least two plausible processes that could account for this: (a) a gradual increase in activity that reaches a peak at orgasm and consequently shows up only after the activity crosses a threshold that was arbitrarily set with the fMRI instrumentation, or (b) a brain region that is specifically related to the unique quality of orgasm that becomes activated only at orgasm. Further research is required to distinguish between these two possible processes.

IMPLICATIONS FOR PRACTITIONERS

The physical and psychological trauma of spinal cord injury produce prolonged and profound disruption in all phases of life and social relationships, among the most troubling of which are bowel and bladder management, mobility, and sexuality. In our interviews of women with spinal cord injury (Komisaruk, Gerdes, & Whipple, 1997; Whipple, Gerdes, & Komisaruk, 1996), most of the women lamented the fact that in their rehabilitation therapy, sexuality received perfunctory discussion, lumped together with limited sessions on bowel and bladder management. Through our interviews, a trajectory of adjustment to the occurrence of the spinal cord injury became evident. We termed the sequence of components of the trajectory as, initially, *cognitive-genital dissociation,* followed by a phase of *sexual disenfranchisement,* and later, *sexual rediscovery* (Richards, Tepper, Whipple, & Komisaruk, 1997). In cognitive-genital dissociation, immediately postinjury, sexuality was of low priority. As one woman described it, "I closed up . . . I shut down every response . . . I shut down being happy . . . I shut everything off." There followed a period of relationship disconnection and disappointment—sexual disenfranchisement. Among 11 women in committed relationships, 8 believed that their spinal cord injury played a role in breaking up their relationships. As one woman said, "It wasn't physically gratifying . . . I didn't orgasm like I did before . . . We had finally gotten to the point of sexual intimacy and it didn't work." Gradually, over a period of years, a phase of sexual exploration emerged—sexual rediscovery. Of 13 women interviewed who experienced orgasms prior to their injury, 8 reported experiencing orgasm at latencies ranging from 2 to 15 years postinjury. One woman said of the need for communication, resourcefulness, and creativity with a partner, "It kind of forces you to be inventive . . . and more experimental." Still later, after 8 to 15 years, the women described becoming more comfortable with their sexuality, developing the ability to communicate with a partner and to be open about their own sexual needs, starting to develop a sense of sexual self-esteem.

In our laboratory studies with the women with spinal cord injury, more than half were surprised that they could feel the vaginal or cervical stimulation and experience pleasurable sensations including orgasms (see Komisaruk, Gerdes, & Whipple, 1997, for quotes of their descriptions). The women commonly said that when they told their physicians that they experience genital sensations, menstrual discomfort, or orgasms, they were told that because of the nature of their injury, it is "impossible," that they are imagining it, or that it is "all in their head." The present findings provide evidence that, contrary to doctrine, a physical, neurologically functional, sensory pathway from the vagina and cervix exists via the Vagus nerves, which can convey sensory activity directly to the brain, bypassing the spinal cord. As a result, although this pathway may function to a greater or lesser extent after spinal cord injury, it is possible that even after a complete upper spinal cord interruption, a woman may experience genital sensation and sexual pleasure. This possibility should not be summarily dismissed, and it may even be advisable to explore the possibility of increasing its salience through rehabilitation.

CONCLUSIONS

The above findings lead us to three major conclusions: First, in women, the Vagus nerves provide a genital (cervical) sensory pathway that completely bypasses the spinal cord, projecting directly to the brain, and consequently can provide genital sensation despite complete spinal cord interruption at any level. This should be heeded by health care practitioners to avoid informing women, after such injury, that genital sensation is no longer possible. Instead, individual testing is indicated before making such categorical assumptions. Second, the genital sensory activity conveyed by the Vagus nerves, to the exclusion of the genitospinal nerves, is evidently adequate to induce orgasm in some women. This provides a minimal and thereby "simplified" informational neural input that could facilitate elucidation of the neural mechanisms underlying orgasm, which is an intriguing and compelling neurophysiological process. Third, specific brain regions appear to be uniquely activated at orgasm. (For further details and discussion, see Komisaruk et al., 2004.)

REFERENCES

American Spinal Injury Association. (1992). *Standards for neurological and functional classification of spinal cord injury—Revised*. Chicago: Author.

Balkin, T. J., Braun, A. R., Wesensten, N. J., Jeffries, K., Varga, M., Baldwin, P., et al. (2002). The process of awakening: A PET study of regional brain activity patterns mediating the re-establishment of alertness and consciousness. *Brain, 125*, 2308–2319.

Berard, E. J. J. (1989). The sexuality of spinal cord injured women: Physiology and pathophysiology—A review. *Paraplegia, 27*, 99–112.

Berkley, K. J., Hotta, H., Robbins, A., & Sato, Y. (1990). Functional properties of afferent fibers supplying reproductive and other pelvic organs in pelvic nerve of female rats. *Journal of Neurophysiology, 63*, 256–272.

Bianca, R., Sansone, G., Cueva-Rolon, R., Gomez, L. E., Ganduglia-Pirovano, M., Beyer, C., et al. (1994). Evidence that the Vagus nerve mediates a response to vaginocervical stimulation after spinal cord transection in the rat. *Society for Neuroscience Abstracts, 20,* 961.

Blaicher, W., Gruber, D., Bieglmayer, C., Blaicher, A. M., Knogler, W., & Huber, J. C. (1999). The role of oxytocin in relation to female sexual arousal. *Gynecologic and Obstetric Investigation, 47,* 125–126.

Bonica, J. J. (1967). *Principles and practices of obstetric analgesia and anesthesia.* Philadelphia: Davis.

Bornhovd, K., Quante, M., Glauche, V., Bromm, B., Weiller, C., & Buchel, C. (2002). Painful stimuli evoke different stimulus–response functions in the amygdala, prefrontal, insula and somatosensory cortex: A single-trial fMRI study. *American Journal of Gastroenterology, 97,* 654–661.

Carmichael, M. S., Humbert, R., Dixen, J., Palmisano, G., Greenleaf, W., & Davidson, J. M. (1987). Plasma oxytocin increases in the human sexual response. *Journal of Clinical Endocrinology and Metabolism, 64,* 27–31.

Carmichael, M. S., Warburton, V. L., Dixen, J., & Davidson, J. M. (1994). Relationships among cardiovascular, muscular, and oxytocin responses during human sexual activity. *Archives of Sexual Behavior 23,* 59–79.

Casey, K. L., Morrow, T. J., Lorenz, J., & Minoshima, S. (2001). Temporal and spatial dynamics of human forebrain activity during heat pain: Analysis by positron emission tomography. *Journal of Neurophysiology, 85,* 951–959.

Cole, T. (1975). Sexuality and physical disabilities. *Archives of Sexual Behavior, 4,* 389–403.

Collins, J. J., Lin, C. E., Berthoud, H. R., & Papka, R. E. (1999). Vagal afferents from the uterus and cervix provide direct connections to the brainstem. *Cell and Tissue Research, 295,* 43–54.

Cross, B. A., & Wakerley, J. B. (1977). The neurohypophysis. *International Review of Physiology, 16,* 1–34.

Cueva-Rolon, R., Sansone, G., Bianca, R., Gomez, L. E., Beyer, C., Whipple, C., et al. (1994). Evidence that the Vagus nerve mediates some effects of vaginocervical stimulation after genital deafferentation in the rat. *Society for Neuroscience Abstracts, 20,* 961.

Cueva-Rolon, R., Sansone, G., Bianca, R., Gomez, L. E., Beyer, C., Whipple, B., & Komisaruk, B. R. (1996). Vagotomy blocks responses to vaginocervical stimulation after genitospinal neurectomy in rats. *Physiology and Behavior, 60,* 19–24.

Cunningham, S. T., Steinman, J. L., Whipple, B., Mayer, A. D., & Komisaruk, B. R. (1991). Differential roles of hypogastric and pelvic nerves in analgesic and motoric effects of vaginocervical stimulation in rats. *Brain Research, 559,* 337–343.

Ding, Y. Q., Shi, J., Wang, D. S., Xu, J. Q., Li, J. L., & Ju, G. (1999). Primary afferent fibers of the pelvic nerve terminate in the gracile nucleus of the rat. *Neuroscience Letters, 272,* 211–214.

Erskine, M. S., & Hanrahan, S. B. (1997). Effects of paced mating on c-fos gene expression in the female rat brain. *Journal of Neuroendocrinology, 9,* 903–912.

Felten, D. L., & Jozefowicz, R. F. (2003). *Netter's atlas of human neuroscience.* Teterboro, NJ: Icon Learning Systems.

Ferguson, J. K. W. (1941). A study of the motility of the intact uterus at term. *Surgical Gynecology and Obstetrics, 73,* 359–366.

Gomora, P., Beyer, C., Gonzalez-Mariscal, G., & Komisaruk, B. R. (1994). Momentary analgesia produced by copulation in female rats. *Brain Research, 656,* 52–58.

Hamann, S. B., Herman, R. A., Nolan, C. L., & Wallen, K (2002). *Sex differences in neural response to visual sexual stimuli revealed by fMRI (Program No. 620.14): Abstract viewer/itinerary planner* [CD-ROM]. Washington, DC: Society for Neuroscience.

Holstege, G., Georgiadis, J. R., Paans, A. M. J., Meiners, L. C., van der Graaf, F. H. C. E., & Reinders, A. A. T. S. (2003). Brain activation during human male ejaculation. *Journal of Neuroscience, 23,* 9185–9193.

Hubscher, C. H., & Berkley, K. J. (1994). Responses of neurons in caudal solitary nucleus of female rats to stimulation of vagina, cervix, uterine horn and colon. *Brain Research, 664,* 1–8.

Hubscher, C. H., & Berkley, K. J. (1995). Spinal and vagal influences on the responses of rat solitary nucleus neurons to stimulation of uterus, cervix and vagina. *Brain Research, 702,* 251–254.

Kettl, P., Zarefoss, S., Jacoby, K., Garman, C., Hulse, C., Rowley, F., et al. (1991). Female sexuality after spinal cord injury. *Sexuality and Disability, 9,* 287–295.

Kirchner, A., Birklein, F., Stefan, H., & Handwerker, H. O. (2000). Left Vagus nerve stimulation suppresses experimentally induced pain. *Neurology, 55,* 167–171.

Komisaruk, B. R., Adler, N. T., & Hutchison, J. (1972, December 22). Genital sensory field: Enlargement by estrogen treatment in female rats. *Science, 178,* 1295–1299.

Komisaruk, B. R., Bianca, R., Sansone, G., Gomez, L. E., Cueva-Rolon, R., Beyer, C., & Whipple, B. (1996) Brain-mediated responses to vaginocervical stimulation in spinal cord-transected rats: Role of the Vagus nerves. *Brain Research, 708,* 128–134.

Komisaruk, B. R., Cueva-Rolon, R., Gomez, L., Ganduglia-Pirovano, M., Sansone, G., & Bianca, R. (1995). Vagal afferent electrical stimulation produces pupil dilatation in the rat. *Society for Neuroscience Abstracts, 21,* 1156.

Komisaruk, B. R., Gerdes, C. A., & Whipple, B. (1997). "Complete" spinal cord injury does not block perceptual responses to genital self-stimulation in women. *Archives of Neurology, 54,* 1513–1520.

Komisaruk, B. R., Mosier, K. M., Criminale, C., Liu, W.-C., Zaborszky, L., Whipple, B., & Kalnin, A. J. (2002). Functional localization of brainstem and cervical spinal cord nuclei in humans with fMRI. *American Journal of Neuroradiology, 23,* 609–607.

Komisaruk, B. R., & Whipple, B. (1984). Evidence that vaginal self-stimulation in women suppresses experimentally-induced finger pain. *Society for Neuroscience Abstracts, 10,* 675.

Komisaruk, B. R., & Whipple, B. (1994). Complete spinal cord injury does not block perceptual responses to vaginal or cervical self-stimulation in women. *Society for Neuroscience Abstracts, 20,* 961.

Komisaruk, B. R., & Whipple, B. (1995). The suppression of pain by genital stimulation in females. *Annual Review of Sex Research, 6,* 151–186

Komisaruk, B. R., & Whipple, B. (1998). Love as sensory stimulation: Physiological consequences of its deprivation and expression. *Psychoneuroendocrinology, 23,* 927–944.

Komisaruk, B. R., & Whipple, B. (2000). How does vaginal stimulation produce pleasure, pain and analgesia? In R. B. Fillingim (Ed.), *Pain research and management: Vol. 17. Sex, gender and pain* (pp. 109–134). Seattle, WA: IASP Press.

Komisaruk, B. R., Whipple, B., Crawford, A., Grimes, S., Kalnin, A. J., Mosier, K., et al. (2002). *Brain activity (fMRI and PET) during orgasm in women in response to vaginocervical self-stimulation (Program No. 841.17): Abstract viewer/itinerary planner* [CD-ROM]. Washington, DC: Society for Neuroscience.

Komisaruk, B. R., Whipple, B., Crawford, A., Grimes, S., Liu, W-C., Kalnin, A., & Mosier, K. (2004). Brain activation during vaginocervical self-stimulation and orgasm in women with complete spinal cord injury: fMRI evidence of mediation by the Vagus nerves. *Brain Research, 1024,* 77–88.

Komisaruk, B. R., Whipple, B., Gerdes, C., Harkness, B., & Keyes, J. W., Jr. (1997). Brainstem responses to cervical self-stimulation: preliminary PET-scan analysis. *Society for Neuroscience Abstracts, 23,* 1001.

Komisaruk, B., Whipple, B., Gerdes, C. A., Sansone, G., Bianca, R., Cueva-Rolon, R., et al. (1996, June). Further evidence that the Vagus nerve conveys genital sensory activity directly to the brain, bypassing the spinal cord. In *Proceedings of the International Behavioral Neuroscience Society.* San Antonio, TX: International Behavioral Neuroscience Society.

Kow, L. M., & Pfaff, D. W. (1973–74). Effects of estrogen treatment on the size of receptive field and response threshold of pudendal nerve in the female rat. *Neuroendocrinology, 13,* 299–313.

Masters, W. H., & Johnson, V. E. (1966). *Human sexual response.* Boston: Little, Brown.

Maixner, W., & Randich, A. (1984). Role of the right vagal nerve trunk in antinociception. *Brain Research, 298,* 374–377.

Ness, T. J., Randich, A., Fillingim, R., Faught, R. E., & Backensto, E. M. (2001). Left vagus nerve stimulation suppresses experimentally induced pain. *Neurology, 56,* 985–986.

Ortega-Villalobos, M., Garcia-Bazan, M., Solano-Flores, L. P., Ninomiya-Alarcon, J. G., Guevara-Guzman, R., & Wayner, M. J. (1990). Vagus nerve afferent and efferent innervation of the rat uterus: An electrophysiological and HRP study. *Brain Research Bulletin, 25,* 365–371.

Peters, L. C., Kristal, M. B., & Komisaruk, B. R. (1987). Sensory innervation of the external and internal genitalia of the female rat. *Brain Research, 408*, 199–204.

Pfaus, J. G., Damsma, G., Wenkstern, D., & Fibiger, H. C. (1995). Sexual activity increases dopamine transmission in the nucleus accumbens and striatum of female rats. *Brain Research, 693*, 21–30.

Pfaus, J. G., & Heeb, M. M. (1997). Implications of immediate-early gene induction in the brain following sexual stimulation of female and male rodents. *Brain Research Bulletin, 44*, 397–407.

Ploner, M., Gross, J., Timmermann, L., & Schnitzler, A. (2002). Cortical representation of first and second pain sensation in humans. *Proceedings of the National Academy of Sciences USA, 99*, 12444–12448.

Randich, A., & Gebhart, G. F. (1992). Vagal afferent modulation of nociception. *Brain Research Reviews, 17*, 77–99.

Richards, E., Tepper, M., Whipple, B., & Komisaruk, B. R. (1997). Women with complete spinal cord injury: A phenomenological study of sexuality and relationship experiences. *Sexuality and Disability, 15*, 271–283.

Rowe, D. W., & Erskine, M. S. (1993). c-Fos proto-oncogene activity induced by mating in the preoptic area, hypothalamus and amygdala in the female rat: Role of afferent input via the pelvic nerve. *Brain Research, 621*, 25–34.

Sansone, G. R., Gerdes, C. A., Steinman, J. L., Winslow, J. T., Ottenweller, J. E., Komisaruk, B. R., & Insel, R. T. (2002). Vaginocervical stimulation releases oxytocin within the spinal cord in rats. *Neuroendocrinology, 75*, 306–315.

Sipski, M. L., & Alexander, C. J. (1995). Spinal cord injury and female sexuality. *Annual Review of Sex Research, 6*, 224–244.

Sipski, M. L., Alexander, C. J., & Rosen, R. C. (1995). Orgasm in women with spinal cord injuries: A laboratory-based assessment. *Archives of Physical Medicine and Rehabilitation, 76*, 1097–1102.

Stein, E. A., Pankiewicz, J., Harsch, H. H., Cho, J. K., Fuller, S. A., Hoffmann, R. G., et al. (1998). Nicotine-induced limbic cortical activation in the human brain: A functional MRI study. *American Journal of Psychiatry, 155*, 1009–1015.

Tetel, M. J., Getzinger, M. J., & Blaustein, J. D. (1993). Fos expression in the rat brain following vaginal-cervical stimulation by mating and manual probing. *Journal of Neuroendocrinology, 5*, 397–404.

Tiihonen, J., Kuikka, J., Kupila, J., Partanen, K., Vainio, P., Airaksinen, J., et al. (1994). Increase in cerebral blood flow of right prefrontal cortex in man during orgasm. *Neuroscience Letters, 170*, 241–243.

Toniolo, M. V., Whipple, B. H., & Komisaruk, B. R. (1987). Spontaneous maternal analgesia during birth in rats. In *Proceedings of the NIH Centennial MBRS-MARC Symposium* (p. 100). Bethesda, MD: National Institutes of Health.

Travers, S. P., & Norgren, R. (1995). Organization of orosensory responses in the nucleus of the solitary tract of rat. *Journal of Neurophysiology, 73*, 2144–2162.

Veening, J. G., & Coolen, L. M. (1998). Neural activation following sexual behavior in the male and female rat brain. *Behavioral Brain Research, 92*, 181–193.

Wersinger, S. R., Baum, M. J., & Erskine, M. S. (1993). Mating-induced FOS-like immunoreactivity in the rat forebrain: A sex comparison and a dimorphic effect of pelvic nerve transection. *Journal of Neuroendocrinology, 5,* 557–568.

Whipple, B. (1990). Female sexuality. In J. Leyson (Ed.), *Sexual rehabilitation of the spinal cord injured patient* (pp 19–38). Clifton, NJ: Humana Press.

Whipple, B., Gerdes, C., & Komisaruk, B. R. (1996). Sexual response to self-stimulation in women with complete spinal cord injury. *Journal of Sex Research, 33,* 231–240.

Whipple, B., & Komisaruk, B. R. (1985). Elevation of pain threshold by vaginal stimulation in women. *Pain, 21,* 357–367.

Whipple, B., & Komisaruk, B. R. (1988). Analgesia produced in women by genital self-stimulation. *Journal of Sex Research, 24,* 130–140.

Whipple, B., & Komisaruk, B. R. (2002). Brain (PET) responses to vaginal-cervical self-stimulation in women with complete spinal cord injury: Preliminary findings. *Journal of Sex and Marital Therapy, 28,* 79–86.

Wildt, L., Kissler, S., Licht, P., & Becker, W. (1998). Sperm transport in the human female genital tract and its modulation by oxytocin as assessed by hysterosalpingoscintigraphy, hysterotonography, electrohysterography and Doppler sonography. *Human Reproduction Update, 4,* 655–666.

6

NEUROANATOMICAL AND NEUROTRANSMITTER DYSFUNCTION AND COMPULSIVE SEXUAL BEHAVIOR

ELI COLEMAN

Individuals with compulsive sexual behavior (CSB) are often unable to control their sexual behavior, may act out impulsively, and/or are often plagued by intrusive obsessive thoughts and driven behaviors. They often perceive their behavior as excessive and experience serious consequences (Coleman, Raymond, & McBean, 2003). It is unfortunate that clinical sexologists appear unable to reach consensus on what to call or how to treat such sexual behavior. Terms have been used to describe this phenomenon, including *hypersexuality, erotomania, nymphomania, satyriasis,* and, most recently, *sexual addiction* and *compulsive sexual behavior.* The terminology has often implied different values, attitudes, and theoretical orientations, and the debate continues regarding classification, causes, and treatment (Coleman, 1990, 1991, 1992). Despite the lack of consensus of terminology, there is consensus that just as there can be problems with hyposexuality, there are individuals who experience hypersexuality (Coleman, 2003).

In my work and throughout this chapter, I use the term *compulsive sexual behavior* (CSB) to describe this syndrome. I use this term to describe the clinical phenomenon but leave open the possibility of multiple etiologies and treatments. Some of the behavioral problems appear to be problems of impulse control rather than driven by obsessive-compulsive mechaynisms. These behaviors are not always ego-dystonic, as in classic obsessive-compulsive disorder (Raymond, Coleman, Benefield, & Miner, 2003). Other times, it seems to be a function of a generalized hypersexual drive, or it seems to function as a dysregulated pleasure-seeking behavior (Coleman et al., 2003).

THE NATURE OF COMPULSIVE SEXUAL BEHAVIOR

Compulsive sexual behaviors can be divided into two main types: paraphilic and nonparaphilic (Coleman, 1991). Paraphilic CSB involves unconventional sexual activities that often interfere in social, interpersonal, and occupational functioning. In the recent *Diagnostic and Statistical Manual of Mental Disorders* (4th ed., text revision; *DSM–IV–TR*), paraphilias are defined as "recurrent, intense sexually arousing fantasies, sexual urges, or behaviors generally involving (1) nonhuman objects, (2) the suffering or humiliation of oneself or one's partner, or (3) children or other non-consenting persons" (American Psychiatric Association, 2000, p. 566). The definition goes on to explain, "The behavior, sexual urges, or fantasies cause clinically significant distress in social, occupational, or other important areas of functioning" (p. 566).

Nonparaphilic CSB involves conventional behaviors. The *DSM–IV–TR* describes one example under the heading of "Sexual Disorder Not Otherwise Specified" as "Distress about a pattern of repeated sexual relationships involving a succession of lovers who are experienced by the individual only as things to be used" (American Psychiatric Association, 2000, p. 582). However, there are a number of other types of nonparaphilic CSB that have been identified (Coleman, 1992).

In the past, my colleagues and I have used a slight variation of the paraphilia diagnostic criteria (Coleman et al., 2003):

A. Intense, sexually arousing behaviors, sexual urges, or fantasies that involve one of the following:
   (1) Compulsive cruising and multiple sexual partners
   (2) Compulsive fixation on an unattainable partner
   (3) Compulsive autoeroticism
   (4) Compulsive use of the Internet for sexual purposes
   (5) Compulsive use of erotica
   (6) Compulsive multiple love relationships
   (7) Compulsive sexuality in a relationship

B. The fantasies, sexual urges, or behaviors cause clinically significant distress or impairment in social, occupational, or other important areas of functioning.

C. Not because of another medical condition, substance use disorder, or attributable to another Axis I or II disorder, developmental disorder. Must take into account norms of gender, sexual orientation, and sociocultural groups.

D. Duration of at least 6 months.

There are myriad hypothesized causes for CSB. However, the focus of this chapter is on neuroanatomical and neurotransmitter dysregulation and dysfunction that have been implicated in causing at least some cases of CSB. Studies are under way exploring hypothesized neuroanatomic and neurophysiological abnormalities through functional neuroimaging. It is hoped that these studies will lead to improved pharmacologic treatment and will be a helpful adjunct for the necessary psychotherapeutic treatment (Coleman, 1995; Coleman, Cesnik, Moore, & Dwyer, 1992).

NEUROANATOMICAL ABNORMALITIES

It has been hypothesized that sexual abnormalities may have their origin in the centers and pathways in the brain that are responsible for sexual arousal and mating behavior. The main part of the brain that has been implicated is the limbic system consisting of the amygdala, the hippocampus, and the hypothalamus. This is also known as the paleomammalian cortex. In addition to sexual functions, the limbic system is also responsible for other elemental functions of individual and species survival, including fight-or-flight responses. The hippocampus and amygdala are covered by the temporal lobe of the neomammalian cortex. The frontal lobe acts as an executive function over the rest of the brain.

Disturbed communication pathways between the frontal and temporal lobe have been implicated in many of the neurological disturbances that may be responsible for CSB. Also, the close proximity of sexual arousal and fight-or-flight responses in the brain is implicated in disorders of sexual violence (Money, 1990). There have been a number of case reports of individuals who developed pedophilic or hypersexual behaviors after some type of cerebral event, such as stroke, traumatic injury, or development of diseases that affect cerebral structure. In addition, disinhibited sexual activity and paraphilias have been reported following lesions in the frontal lobe, hypothalamus, and septum (Frohman, Frohman, & Moreault, 2002; Miller, Cummings, McIntyre, Ebers, & Grode, 1986).

Most of the brain abnormalities have been found in paraphilic CSB. Berlin (1988) and Langevin (1990) have provided good reviews of this literature. They have cautioned, however, that *as a group* paraphilics do not differ from control groups on the basis of brain abnormalities. However, medical technology is advancing so fast that one is now able to see and understand brain mechanisms and neurotransmitter pathways in ways that previously were not possible. The use of computer tomography imaging (CT), magnetic resonance imaging (MRI), functional magnetic resonance imaging (fMRI), and positron emission tomography (PET) have made revolutionary inroads in the understanding of psychiatric disorders and the development of pharmacological treatments. The search for neuroanatomical and neurophysiological correlates of sexual disorders will obviously increase in intensity in the near future. It is naive to think that any simple brain mechanism or neurotransmitter will explain the complexity of human sexual expression or disorders. However, there should be great optimism of further unlocking the mysteries of the brain, neurochemistry, and its relationship to sexual behavior.

Frontal Lobe Abnormalities

Frontal lobe dysfunction can lead to disinhibition of sexual behavior and hypersexual behavior or CSB. Based on a review of the literature, Miller et al. (1986) concluded that frontal lobe abnormalities were more associated with hypersexuality and that the temporal lobe dysfunctions were more related to pedophilic sexual interests.

The frontal lobes are responsible for "executive functioning" such as response inhibition, planning, and verbal mediation of behavior. Lesions of frontal brain have been shown to produce impulse control problems, including hypersexual behavior (Damasio, Tranel, & Damasio, 1990; Elliot & Biever, 1996).

Studies are under way to examine structural and functional MRI scans in an attempt to understand what portions of frontal lobe as well as other parts of the brain are functioning abnormally in individuals with CSB. It is hoped that these studies will further unlock the mysteries of the potential neuroanatomical and neurophysiological factors involved in CSB and lead to improved treatment.

Temporal Lobe Abnormalities

There has been a great deal of focus on the hypothesized involvement of the temporal lobe in CSB. Temporal lobe abnormalities, which have been associated with hypersexuality, also may be involved in development of various paraphilias such as fetishes and pedophilia. John Money has noted that, in some case reports, comorbidity exists between paraphilias with temporal lobe lesion(s) and seizures. In these patients, when the lesion can be corrected through neurosurgery, the paraphilia disappears (Money & Lamarz, 1990). In addition, temporal lobe epilepsy has been associated with states that are a common feature of CSB (Mesulam, 1981). For example, Money and Lamarz (1990) described the fuguelike state, which is an altered state of consciousness and which is activated when the paraphilia engage in their paraphilic activity. They noted that there may be a correlation with some temporal lobe seizurelike activity. In some cases, disturbances of cortical electrical activity in the regions of the brain related to the limbic system may provide explanation for obsessive thoughts, or fuguelike states associated with CSB, or the compulsive behaviors themselves.

Further evidence of temporal lobe involvement comes from studies of stroke victims. Although hyposexuality is a common problem in stroke patients, some clinicians have noted hypersexuality and unusual sexual behavior among patients after a stroke (Monga, Monga, Raina, & Hardjasudarma, 1986). In the patients who showed hypersexuality, CT scans indicated temporal lobe lesions, and all had a history of poststroke seizure activity.

Faulty functioning could be triggered by the growth of a tumor or from an open or closed head injury. This can result in seizurelike activity of the nonconvulsive type. An example of this would be the Kluver-Bucy syndrome, a disorder in which a temporal lobe lesion leads to hypersexuality. This was first noted in animals (Kluver & Bucy, 1939) and then in humans (Goscinski, Kwiatkowski, Polak, Orlowiejska, & Partyk, 1997; Nakada, Lee, Kwee, & Lerner, 1984; Pearce & Miller, 1973). "Hypersexuality" and alterations in sexual orientation have been noted in a number of patients following brain injury (Elliott & Biever, 1996; Huws, Shubsachs, & Taylor, 1991; Mendez, Chow, Ringman, Twitchell, & Hinkin, 2000; Miller et al., 1986; Ortego, Miller, Itabashi, & Cummings, 1993).

Researchers have also suggested, on the basis of case studies, that pedophilic interests are associated with abnormalities in the temporal lobe or midbrain adjacent to the hypothalamus (Kolarsky, Madlafousek, & Novotna, 1978; Miller et al., 1986; Ortego et al., 1993). Others have concluded that these temporal lobe disturbances actually result in hypersexuality, and this hypersexuality may unmask a previously hidden orientation toward children (Mendez et al., 2000). It should be noted that these findings are based on very few case studies. And, the majority of individuals with temporal lobe epilepsy or temporal lobe damage (e.g., stroke) or those with temporal lobectomy experience hyposexual desire, but a small percentage of patients experience hypersexuality (Blumer, 1970; Blumer & Walker, 1967; Monga et al., 1986). The reasons for this are unclear.

Although birth defects, head traumas, strokes, or brain surgeries can alter temporal lobe functioning, one can also speculate about the impact of physical or emotional abuse. Individuals with CSB have reported disproportionately high levels of physical and emotional abuse as children (Coleman, 1991). This is true of many psychiatric disorders. Although most patients with CSB do not have temporal lobe epilepsy or known neurological disorders, one speculates what effect physical or emotional abuse or specifically head injuries associated with physical abuse have had on their developing brains (Coleman, 1991).

Some of my patients have suspected temporal lobe disturbances, but neurological testing had been inconclusive. However, this does not rule out that neurological abnormalities exist in these patients. In one case of a pedophile, neurological findings were negative. His inability to concentrate, tendency to lose track of his conversation, cognitive distortions, hyperreactivity, and rapid shifts in personality all pointed to some neurological disturbance. Given this speculation and the fact that he was not improving with psychotherapy, our clinical team decided to give him a trial of a mood stabilizer (carbamazapine). He reported improvement in his cognitive functioning, hyperreactivity, and interpersonal communication. I could observe these behavioral changes, and they were also noted by his therapy group members. In questioning his sister on whether she had noted any changes since her brother began the medication, she gave me an immediate example, "I feel like I just had my first real conversation with him while we were waiting in the reception area." There was no change in his sexual arousal patterns, but his improved cognitive ability and interpersonal skills seemed to be having an effect on his overall self-esteem and control of his pedophilic interests. Therapy was also able to progress at a more rapid pace, and he eventually successfully completed treatment. One can only speculate on the temporal lobe involvement in this case; however, there may have been some involvement as part of the constellation of factors that may have contributed, maintained, or made rehabilitation difficult. Although this is just a case example, I have seen a number of similar cases with positive results.

NEUROPHYSIOCHEMICAL ABNORMALITIES

Chemical reactions in the nervous system, in the form of hormones and other neurochemicals, have significant effects on behavior. In this section I discuss some ways that abnormalities in these processes might affect CSB.

Testosterone

Studies that have tried to correlate testosterone and CSB focused on sexual offenders and have been fraught with methodological problems and contradictory results. Not all sexual offenders necessarily meet diagnostic criteria for paraphilic disorder. In addition, basic understanding of endocrinology information is often ignored in interpreting the findings of many of these studies. For example, clinically relevant hormonal mechanisms are rarely reflected in total plasma testosterone levels. Measurement of the unbound plasma testosterone is rarely measured, and this is the biologically active portion of the total. The range of normal plasma testosterone levels is large. The clinically relevant information is probably more reflected in receptor sites; however, there have been no easy means developed to measure hormone receptivity. Thus, most studies continue to measure total plasma testosterone without careful interpretation of this information.

In one study, plasma testosterone levels in 52 male rapists and 12 male child molesters were measured (Rada, Laws, & Kellner, 1976). The ranges of the rapists and the child molesters were within normal limits. However, the rapists whose crimes were judged to be most violent had significantly higher plasma testosterone levels than other rapists and child molesters. Rada et al. were careful to point out the limitations in interpreting these results.

Researchers have noted that external factors could account for these variations in plasma testosterone levels (Rada, Kellner, & Winslow, 1976). For example, the stresses of having been arrested or incarcerated could alter and influence plasma testosterone. Changes could also be due to chronic ingestion of alcohol, which has been noted to reduce plasma testosterone levels (Mendelson & Mello, 1974). The effects of stress and anxiety must be taken into account when interpreting data on plasma testosterone levels. Even methods of gathering plasma testosterone levels (time and method) can create spurious differences. It may be that the precision of assay techniques was still too crude to note any abnormalities.

However, in assessing the role of sexual hormones on sexual behavior in 1977, John Bancroft was not hopeful about science providing some clear answers in the near future. He suggested that "those psychiatrists looking for simple biological explanations of human behavior would be advised to search elsewhere" (Bancroft, 1977, p. 555). Researchers have essentially followed this advice in terms of looking for abnormalities of testosterone studies. Since the initial inconclusive studies, this line of research was abandoned. However, with better assay techniques and more precision as to type of offender, it seems that it would be useful to conduct further studies of plasma testosterone levels, recognizing the probable complexity of factors associated with offending behavior.

Although studies have not been conclusive about abnormalities of testosterone levels, the efficacy of antiandrogen treatment for sex offenders and individuals with CSB has been fairly well documented (Bradford, 2000). It is assumed that lowering testosterone levels reduces libido and gives greater control to individuals who feel "driven" by their paraphilic or nonparaphilic urges. However, because lowering testosterone has a cascading effect on other neurotransmitters, there may be a more complex explanation. Because I have always felt that CSB is driven more by anxiety reduction mechanisms than by sexual drive, I have been more interested in the neurophysiochemical processes involved in stress and the pharmacological treatments that could alleviate a person's maladaptive stress response.

Autonomic Reactivity and Neurotransmitter Dysregulation

Because many individuals with CSB have experienced abuse in childhood, it is interesting to speculate what effect this might have had on their neurophysiochemical functioning. There are no data on individuals with CSB in this regard, but there is a body of research on individuals with posttraumatic stress disorder (PTSD) that might suggest similarities. Studies in animals and humans have focused on the specific brain areas and circuits involved in anxiety and fear, which are important for understanding anxiety disorders such as PTSD. The fear response is coordinated by a small structure deep inside the brain, called the amygdala. It has been suggested that the different anxiety disorders are associated with abnormal activation of the amygdala (LeDoux, 1998).

People with PTSD tend to have abnormal levels of key hormones involved in response to stress (Yehuda, 1998). Individuals who have PTSD have been shown to display increased autonomic reactivity to auditory stimuli in laboratory conditions (Kolb & Mutalipassi, 1982). van der Kolk (1983) described how patients with PTSD have a poor tolerance for arousal, and this condition interferes with cognitive responses to stressful stimuli. He suggested that these patients experience a conditioned response to emotional and sensory stimulation and react with psychomotor discharge or by psychological withdrawal. Their overt behavior may be regulated by conditioned autonomic reactivity. Because many individuals with CSB have been traumatized in childhood, this could result in increases in autonomic reactivity as well. Behaviors (including CSB) may be viewed as an inappropriate and exaggerated response to a perceived threat.

These neurophysiological abnormalities may possibly have a genetic basis, which gives the individual an inborn predisposition for the development of these neurophysiological abnormalities causing increased autonomic reactivity. Parents may pass on this genetic trait to their children. This may explain why some individuals who experience trauma develop maladaptive stress responses whereas others have more resilience. For the more vulnerable or less resilient, events or traumas that create hormonal abnormalities at critical events of the development of the brain may later influence psychosexual development and functioning.

This line of research on PTSD has implications for understanding some of the behavior of persons with PTSD and therapy. Psychotherapy can trigger exaggerated emotional responses, over which the individual has little or no control. The overreactivity can be easily misinterpreted by the therapist. The client may be seen as unmanageable, resistant, or avoidant. van der Kolk (1984) indicated the importance of using psychopharmacologic agents to focus on these patients' hyperreactivity, which might include anxiety, impulsivity, anger outbursts, intrusive nightmares or flashbacks, insomnia, depression, and anhedonia.

van der Kolk (1984) provided another possible explanation for the repetitive nature of CSB. There are autonomic nervous system explanations that provide an alternative to the common psychodynamic view of etiology of CSB. The psychodynamic view postulated that the individual was in some ways trying to resolve conflicts created in childhood through coping mechanisms, attempts at mastery, avoidance of anxiety, or to "triumph over the trauma" (Stoller, 1975). Evidence suggests that as a result of trauma, there is altered autonomic nervous system function (Kolb, Burris, & Griffiths, 1984). It has been hypothesized that the trauma produces effects similar to those of inescapable shock. The effects of inescapable shock on the central neurotransmitter systems have been studied extensively. In animal studies, it is known that exposure to inescapable shock increases norepinephrine turnover, which results in depleted brain norepinephrine; increases plasma catecholamine levels; decreases brain dopamine and serotonin; and increases acetylcholine (van der Kolk, Greenberg, Boyd, & Krystal, 1985). Further exposure to stress (even minor) will evoke a maladaptive stress response. van der Kolk (1988) noted that child abuse is correlated with neurochemical alterations. It may be that such fearful experiences are recorded in one's nervous system. A conditioned fearful response may be then triggered neurophysiologically by any fearful stimulus (see Figure 6.1).

Dysregulation of serotonin, dopamine, and norepinephrine could result in an abnormal stress response but could be directly implicated in heightened or inhibited sexual arousal. The brain's neurotransmitter substances, dopamine and serotonin, seem to have particular excitatory and inhibitory functions on sexual drive. Though their exact role is yet to be discovered, it is generally known that dopamine activates erotosexual interest and serotonin acts as an inhibitor (Bitran & Hull, 1987; Gessa & Tagliamonte, 1974). Most research has focused on the roles of serotonin and dopamine.

Dysregulaton of Norepinephrine

Norepinephrine is a neurotransmitter released during stress. One of its functions is to activate the hippocampus, the brain structure involved

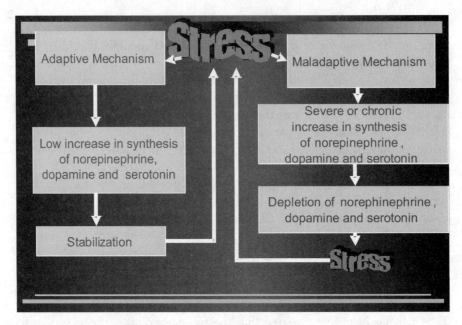

Figure 6.1. Adaptive and maladaptive neurotransmitter response to stress.

with organizing and storing information for long-term memory. Under the extreme stress of trauma, norepinephrine may act longer or more intensely on the hippocampus, leading to the formation of abnormally strong memories that are then experienced as flashbacks or intrusions. It is also interesting to note that individuals with PTSD have shown atrophy or deterioration of the hippocampus (R. J. Davidson, Jackson, & Kalin, 2000). This may explain the development of PTSD symptoms in individuals.

Dysregulation of Serotonin and Dopamine

There is an extensive literature that explains the role of serotonin on sexual arousal and function (Bancila et al., 1999; Marson & McKenna, 1992; Steinbusch, 1981). Animal studies have shown that decreased serotonin levels will enhance sexual arousal and function, whereas increased levels will inhibit sexual function (Marson & McKenna, 1992). In humans, the effect on sexual functioning can be clearly seen in the administration of selective serotonin reuptake inhibitors (SSRIs). These medications were developed for the treatment of depression and were designed to increase serotonin levels. The most common side effect of the SSRIs is the inhibition of ejaculation and orgasm and the decrease in sexual desire (Lane, 1997; Modell, Katholi, Modell, & DePalma, 1997). Although these drugs apparently affect serotonin levels, their pharmacology is vastly more complex, and it is assumed that other neurotransmitter systems are being directly or indirectly affected. It is known, for example, that increasing serotonin will have cascading effects on other neurotransmitter systems. In terms of CSB, serotonergic agents have been found to be effective in treating this clinical syndrome (Coleman, 1991; Federoff, 1993; Kafka & Coleman, 1991).

Dopamine also plays an important function in sexual arousal. Numerous pharmacological studies in animals and humans support this conclusion. In general, dopaminergic agonists facilitate several aspects of sexual function, including erection, ejaculation, female sexual arousal, and sexual motivation. In addition, dopaminergic agents used for the treatment of Parkinsonism sometimes induce increases in sexual desire. In anecdotal reports, cocaine and amphetamine, which increase dopamine, have been shown to have aphrodisiac qualities. However, it is well known that chronic use of stimulants will result in severe sexual dysfunction, including anorgasmia and lack of sexual desire.

Opioids

Still very little is known about the morphinelike neurotransmitters, the endorphins and enkephalines, and the beta-lipotropin in the pituitary gland (Money, 1980). Endogenous opiates are involved centrally and peripherally. Endorphin activity has been found in many areas of the brain. The enkephalins have been found in exceptionally high proportions in limbic structures and in the hypothalamus, areas involved with emotionality, drive, and stress responses. Endorphins have also been found peripherally in human semen, in the prostate, and in the seminal vesicles. Opiates depress sexual response and delay ejaculation in rats by inhibiting the contractions of the vas deferens and seminal vesicles. Their central action is assumed to occur through the hypothalamic-pituitary-testicular axis by inhibiting luteinizing hormone-releasing factor. However, it has been suggested that they may also act through the dopamine-serotonin pathways. The endogenous opiates are released during sexual arousal and are, at least, partially responsible for the sexual anesthesia and fuguelike state during sexual excitement. It has long been known that opiates will enhance sexual arousal. It has also been known that opium antagonists block many of these actions (McIntosh, Vallano, & Barfield, 1980; Murphy, 1981)

Many drug addicts have used opiates as a form of self-medication for mental disorders (Verebey, Volavka, & Clouet, 1978). Opiates have been reported to act in a way that reduces anxiety, rage, aggression, paranoia, feelings of inadequacy, depression, and suicidal ideation (Khantzian, Mack, & Schatzberg, 1974). When patients have been withdrawn from opiates, many of these symptoms are manifested and then, again, controlled by treating them with methadone. Opiates probably do not have direct control over these symptoms but may reduce stress, which then alters many of the other neurotransmitters involved in these emotional disorders. Chronic use of opiates is associated with severe sexual dysfunction. This suggests that opiates play a central role in regulation of sexual function (Crowley & Simpson, 1978; Cushman, 1972).

When people are in danger, they produce high levels of natural opiates, which can temporarily mask pain. It was found that people with PTSD continue to produce those higher levels even after the danger has passed. It has been suggested this may lead to the blunted emotions associated with the condition (Yehuda, 1998).

It is theorized that a certain level of endorphin activity is needed for normal psychological homeostasis. Altering endorphin levels through exogenous opiates or through other medications, which indirectly affect endorphin activity, may have therapeutic effects in some mental disorders. It must be remembered that the inadequate level of endorphin activity in some individuals may be due to the lack of endorphins that are released or through problems with the receptors. It must also be remembered that the endorphin system is only one of the active peptide systems in the central nervous system. Behavior results from a complex interaction of the endorphin system and all of the other peptide systems, including the well-known neurotransmitter system. An imbalance in one of these systems or an alteration through exogenous means may have a cascading effect on all the systems that regulate mechanisms of the brain and, in turn, affect behavior. As a result, attributing an effect in behavior to a deficiency or alteration of one peptide mechanism is a gross oversimplification.

As illustrated in Figure 6.2, van der Kolk (1984) speculated that neurotransmitter dysregulation explains repetition of traumaticlike experiences. Reexposure to traumatic situations in humans evokes a central nervous system opioid response. This response is analogous to that which is seen in animals exposed to low-level shock after having been exposed to inescapable shock. As a result of the endogenous release of opioid peptides, individuals experience a sense of calm. As indicated earlier, opioid users describe psychoactive effects such as reduction in anxiety, rage, aggression, paranoia, feelings of anxiety, and depression. However, following release of opiate peptides, withdrawal symptoms ensue. It is interesting to note that the symptoms of PTSD parallel symptoms of opiate withdrawal. These common symptoms include anxiety, outbursts of anger or aggressive behavior, sleep disturbances, and hyperalertness. van der Kolk (1983) suggested that these seemingly unrelated syndromes of opiate withdrawal and PTSD may involve common neurological mechanisms that would include a decrease in brain opioid receptor binding associated with central noradrenergic hyperactivity. So, in the case of PTSD, reexposure to a traumatic situation would cause an

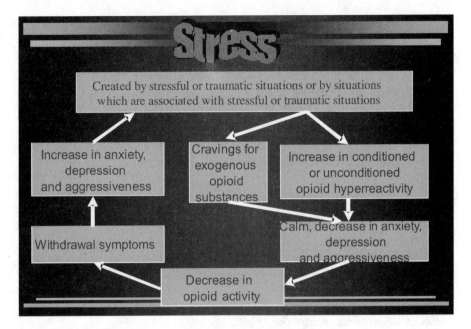

Figure 6.2. Stress and its relationship to opioid activity.

endorphin response, which would give a paradoxical sense of calm and reduction in anxiety, followed by physiological symptoms of opiate withdrawal (increased anxiety, hyperreactivity, and outbursts of aggression).

Again, reexposure to a traumatic situation would relieve these symptoms through release of opiate peptides. This cycle might explain the biological drive to reexperience traumatic events. Or, in the case of obsessive–compulsive disorder, this process may explain the senseless return to behaviors that are often perceived to be unpleasant but create momentary calm when one is engaging in these repetitive behaviors. So, this may be part of the neurophysiochemical explanation of the perpetuating vicious cycle of CSB.

Interrupting this vicious cycle is extremely difficult. This biological process may explain why some of the abreaction techniques of psychotherapy may not be helpful in treating trauma victims. As van der Kolk and colleagues stated:

> If reliving the trauma is followed by a conditioned endorphin release and subsequent withdrawal hyperreactivity, bringing back memories of the trauma in a psychotherapeutic setting might actually increase trauma-seeking behavior, as well as precipitate anxiety and explosive outbursts of anger. Uncovering psychotherapy would thus lead to clinical deterioration. (van der Kolk et al., 1985, p. 321)

Methods to reduce autonomic hyperreactivity may be an important bridge to recalling trauma and resolving the conflict without the individual reexperiencing the trauma and evoking the vicious cycle of the stress response (van der Kolk, McFarlane, & Weisaeth, 1996). Various pharmacotherapies can be powerful tools in reducing autonomic hyperreactivity (J. R. T. Davidson & van der Kolk, 1996). Although cognitive and behavioral psychotherapies and relaxation training designed to modify the conditioned response could be effective (Rothbaum & Foa, 1996), one should recognize the powerful and, oftentimes, resistant-to-change conditioned physiological response.

PHARMACOLOGICAL TREATMENT

There are a number of pharmacologic treatments of CSB that have proved effective in clinical case studies (Coleman et al., 2003; Raymond et al., 2003).

Selective Serotonin Reuptake Inhibitors

Most of the recent attention to the treatment of CSB has been on the effectiveness of the SSRIs. Given the efficacy of SSRIs in the treatment of depression, anxiety, obsessive–compulsive disorder, and impulse-control disorders, when the SSRIs came on the market in the late 1980s, a number of open-label studies have indicated that the SSRIs may be efficacious in the treatment of CSB (Bradford & Gratzer, 1995; Coleman et al., 1992; Federoff, 1993; Kafka, 1991, 1994; Kafka & Prentky, 1992; Stein et al., 1992).

These antidepressants that selectively act on serotonin levels in the brain are effective in reducing sexual obsession and compulsions and their associated levels of anxiety and depression. These medications appear to interrupt the obsessive thinking and help patients control urges to engage in CSB. They also help patients use therapy more effectively (Coleman et al., 1992).

Because of the extensive case reports in the literature, clinicians often choose these medications because of their wide efficacy in treating CSB but also the very common psychiatric comorbidity (Black, Kehrberg, Flumerfelt, & Schlosser, 1997; Raymond, Coleman, Ohlerking, Christenson, & Miner, 1999). The SSRIs have become the frontline medications for treating CSB (Bradford, 2000).

Atypical Antidepresants

In my clinical experience, many patients with nonparaphilic CSB find SSRIs helpful in controlling their sexual obsessions and compulsions but have complained about the sexual side effects. These side effects are particularly troublesome when the patients are trying to reestablish healthy sexual and intimate functioning. Sexual side effects have often been cited as a reason for patients discontinuing their SSRI medications, leading to a return of sexual obsessions and compulsions.

Therefore, researchers began to shift a number of patients to nefazadone. Nefazadone has phenylpiperazine reuptake inhibitor effects and is a competitive antagonist at the serotonin 5 HT_2 receptor subtype. Antagonism of 5 HT_2 receptors has been associated with a lower incidence of sexual side effects in comparison with standard SSRIs (Feiger, Kiev, Shrivastava, Wisselink, & Wilcox, 1996). In a small retrospective study, it was found that most of the patients (11 out of 14) who were switched from an SSRI to long-term nefazodone therapy remained in reported good control or in remission of their sexual obsessions and compulsions (Coleman, Gratzer, Nescvacil, & Raymond, 2000).

Mood Stabilizers

Mood stabilizers can be used when appropriate. In fact, before the advent of the SSRIs, my colleagues and I often used low doses of lithium carbonate to treat cases of CSB. In one of our first published studies, we successfully treated an individual with autoerotic asphyxia (Cesnik & Coleman, 1989). Although there was no evidence of cyclical mood disturbances, this patient had dysthymia along with his self-destructive sexual behavior. We did not consider antiandrogens because we felt that the underlying depressive disorder was a key part of his ritualized behavior. Because most of the older tricyclic antidepressants had been ineffective for treating this type of problem in the past, my colleague had some clinical experience of effectively using lithium carbonate with paraphilic behavior. Thus, we decided on a trial of lithium carbonate. Although we did not completely understand why the lithium carbonate was working, we began to speculate that the serotonergic effects of this medication might be pharmacologic action as well as lithium carbonate's known effectiveness in stabilizing cyclic mood changes and decreasing aggressive and impulsive actions. This was a significant case, as it opened the door to recognizing the importance of using pharmacological treatment along with psychotherapy for a number of cases of CSB. After the success of this case, we began to use this more frequently until the advent of the SSRIs. Upon their arrival, we found the SSRIs to be as effective and easier to administer. We began switching patients to SSRIs with effective results (Coleman et al., 1992). However, for some, the SSRIs were not sufficient in controlling their CSB, and we began to use combinations of SSRIs and lithium carbonate and then other mood stabilizers such as carbamazapine and valproic acid.

Antiandrogens

Medications that suppress the production of testosterone (antiandrogens) have also been used to treat a variety of paraphilic CSBs. Before the advent of the SSRIs, these medications were the most common medications used for treating CSB. The use of antiandrogens, such as cyproterone acetate and medroxyprogesterone acetate, in the treatment of sex offenders was first reported in 1968 (Money, 1968, 1970). Antiandrogen therapy essentially replaced castration as a method of reducing testosterone because of its relative reversibility and the ethical concerns involved. The antiandrogens were problematic because of a variety of potential side effects, such as liver disease, gynecomastia, thrombophlebitis, hyperglycemia, and gall bladder disease. However, with the advent of the SSRIs, the use of antiandrogens waned. SSRIs held an advantage over antiandrogens not only because of their lower side-effect profile but because of their effectiveness in controlling sexual obsessions and compulsions, their ease of administration, and their ability to effectively treat many of the common comorbid psychiatric disorders, especially anxiety and depressive disorders (Raymond et al., 1999).

There has been, however, a resurgence in the interest of antiandrogens with the use of luprolide acetate. Leuprolide acetate is a synthetic analog of leutinizing hormone-releasing hormone (LHRH), which is one of the gonadotropin-releasing hormones. Therefore leuprolide acetate is a potent LHRH agonist. Initially leuprolide acetate stimulates the release of production of testosterone and other testicular steroids, but with chronic administration testicular steroidogenesis is suppressed. Luprolide acetate has been reported to cause fewer side effects, although there still is a concern about osteopenia (Thibaut, Cordier, & Kuhn, 1993, 1996). Apart from the aforementioned studies, there have been a number of other reports, including in our own clinic, which have had successful outcomes (Dickey, 1992; Raymond, Robinson, Kraft, Rittberg, & Coleman, 2002).

Naltrexone

There has been a recent interest in the use of naltrexone in the treatment of CSB. This stems from the discovery of its effectiveness in treating a variety of urge-driven disorders, such as pathological gambling disorder (Kim & Grant, 2001; Kim, Grant, Adson, & Shin, 2001), kleptomania (Grant & Kim, 2002), alcoholism (Volpicelli, Alterman, Hayashida, & O'Brien, 1992), borderline personality disorder with self-injurious behavior (Sonne, Rubey, & Brady, 1996), and eating disorders (e.g., Marrazzi, Bacon, Kinzie, & Luby, 1995). My colleagues and I reported on the first successful treatment of two individuals with CSB with naltrexone (Raymond, Grant, Kim, & Coleman, 2002). Both individuals had severe CSB symptoms and had only shown mild improvements with SSRIs and other antidepressants. When naltrexone was added to the SSRI treatment, the patients lost urges to engage in their prior problematic behavior and were able to stop their CSB.

One can only speculate in these cases on the pharmacologic action. However, if the patients had a dysregulation of the opioid system, the use of naltrexone, an opioid antagonist, would make sense. However, these results are obviously preliminary, and there have not been many other patients who have responded positively in this fashion.

There is an obvious need for controlled clinical trials to develop a more evidenced-based clinical approach to pharmacologic treatment. However, these case studies illustrate the need for careful consideration of the array of treatment options, careful monitoring of CSB symptoms, and the willingness of the physician and patient to try different options when a certain medication is failing. This is not to say that there is a pharmacologic panacea out there for every case of CSB; however, it does illustrate that sometimes the right medication or combination of medications will help resolve the symptoms.

Pharmacological Therapy as an Adjunct to Psychotherapy

As a consequence, pharmacological agents can be extremely helpful in interrupting the vicious cycle of CSB. The psychotherapist may need an important ally in these agents as he or she helps the individual to engage in the stressful experience of psychotherapy, as well as to gain more rapid control of CSB. In psychotherapy, individuals with CSB are often asked to recall their childhood abusive experiences. This reexperiencing of the trauma initially evokes a surprisingly calm or detached response. This is often followed by increased anxiety, which does not seem to be necessarily triggered by any particular event. Or, an emotional response to a particular event (e.g., rejection) is felt far more acutely than the situation would warrant.

As clinicians, we are often puzzled by the repeated or compulsive behaviors that ultimately produce far more negative feelings than the brief temporary relief that a CSB can produce. These neurophysiological mechanisms might help explain how the psychodynamic drives are also regulated by neurochemical changes that occur in the brain (increase in synthesis providing psychic relief followed by increased emotional distress).

CONCLUSION

Pharmacological treatments that address neuroanatomical abnormalities, disturbances, or neurotransmitter dysregulation can be very helpful in assisting individuals with CSB. These disturbances may have been a result of congenital abnormalities, inherited predispositions, or trauma. Physical or psychic trauma can lead to a variety of neuroanatomical or neurophysiological disturbances. No simple brain mechanism or neurotransmitter dysfunction or dysregulation will explain the complexity of human sexual expression or disorders. However, there should be great optimism for further understanding the neurophysiological abnormalities and their relationship to sexual behavior.

There is now an array of pharmacotherapies that can address a variety of possible dysfunctions. Individual differences in neurotransmitter activity because of congenital or environmental influences may explain an individual's neurotransmitter dysregulation and why some medications are more helpful to some than to others. However, pinpointing the cause or the cure is further complicated by the fact that there are complex interactions of stress, individual responsivity, environmental factors, neurotransmitters, and receptors that account for a person's behavior.

Physiological explanations of the CSB are inadequate to fully understand the etiology. However, these physiological explanations may help explain the biological underpinnings of the psychological theories of CSB. In addition, these physiological processes also have important implications for psychotherapy. As a result, the clinician might find pharmacotherapy extremely helpful in treating individuals with CSB, not only in gaining rapid control of CSB but also in helping the patient be more responsive to the rigors and stressors of psychotherapy.

REFERENCES

American Psychiatric Association. (2000). *Diagnostic and statistical manual of mental disorders* (4th ed., text revision). Washington, DC: Author.

Bancila, M., Vergé, D., Rampin, O., Backstrom, J. R., Sanders-Bush, E., McKenna, K. E., et al. (1999). 5-Hydroxytryptamine 2C receptors on spinal neurons controlling penile erection in the rat. *Neuroscience, 92,* 1523–1537.

Bancroft, J. (1977). Hormones and sexual behaviour. *Psychological Medicine, 7,* 553–556.

Berlin, F. (1988). Issues in the exploration of biological factors contributing to the etiology of the "sex offender," plus some ethical considerations. *Annals of the New York Academy of Sciences, 528,* 183–192.

Bitran, D., & Hull, E. M. (1987). Pharmacological analysis of male rat sexual behavior. *Neuroscience and Biobehavioral Review, 11,* 365–389.

Black, D. W., Kehrberg, L. L., Flumerfelt, D. L., & Schlosser, S. S. (1997). Characteristics of 36 subjects reporting compulsive sexual behavior. *American Journal of Psychiatry, 154,* 243–249.

Blumer, D. (1970). Hypersexual episodes in temporal lobe epilepsy. *American Journal of Psychiatry, 126,* 1099–1106.

Blumer, D., & Walker, A. E. (1967). Sexual behavior in temporal lobe epilepsy. *Archives of Neurology, 16*(1), 37–43.

Bradford, J. (2000). Treatment of sexual deviation using a pharmacologic approach. *Journal of Sex Research, 37,* 248–257.

Bradford, J. M. W., & Gratzer, T. G. (1995). A treatment for impulse control disorders and paraphilia: A case report. *Canadian Journal of Psychiatry, 40,* 4–5.

Cesnik, J. A., & Coleman, E. (1989). Use of lithium carbonate in the treatment of autoerotic asphyxia. *American Journal of Psychotherapy, 43,* 277–286.

Coleman, E. (1990). The obsessive-compulsive model for describing compulsive sexual behavior. *American Journal of Preventative Psychiatry and Neurology, 2*(3), 9–14.

Coleman, E. (1991). Compulsive sexual behavior: New concepts and treatments. *Journal of Psychology and Human Sexuality, 4,* 37–52.

Coleman, E. (1992). Is your patient suffering from compulsive sexual behavior? *Psychiatric Annals, 22,* 320–325.

Coleman, E. (1995). Treatment of compulsive sexual behavior. In R. C. Rosen & S. R. Leiblum (Eds.), *Case studies in sex therapy* (pp. 333–349). New York: Guilford Press.

Coleman, E. (2003). Compulsive sexual behavior: What to call it, how to treat it? *SIECUS Report, 31*(5), 12–16.

Coleman, E., Cesnik, J., Moore, A., & Dwyer, S. M. (1992). An exploratory study of the role of psychotropic medications in the treatment of sex offenders. *Journal of Offender Rehabilitation, 18*(3/4), 75–88.

Coleman, E., Gratzer, T., Nescvacil, L., & Raymond, N. (2000). Nefazodone and the treatment of nonparaphilic compulsive sexual behavior: A retrospective study. *Journal of Clinical Psychiatry, 61,* 282–284.

Coleman, E., Raymond, N., & McBean, A. (2003). Assessment and treatment of compulsive sexual behavior. *Minnesota Medicine, 86*(7), 42–47.

Crowley, T. J., & Simpson, A. (1978) Methadone dose and human sexual behavior. *International Journal of Addiction, 31,* 285–295.

Cushman, P. (1972). Sexual behavior in heroin addiction and methadone maintenance: Correlation with plasma luteinizing hormone. *New York State Journal of Medicine, 72,* 1261–1265.

Damasio, A. R., Tranel, D., & Damasio, H. (1990). Individuals with sociopathic behavior caused by frontal damage fail to respond autonomically to social stimuli. *Behavioural Brain Research, 41,* 81–94.

Davidson, J. R. T., & van der Kolk, B. A. (1996). The psychopharmacological treatment of posttraumatic stress disorder. In B. A. van der Kolk, A. C. McFarlane, & L. Weisaeth (Eds.), *Traumatic stress: The effects of overwhelming experience on mind, body and society* (pp. 510–524). New York: Guilford Press.

Davidson, R. J., Jackson, D. C., & Kalin, N. H. (2000). Emotion, plasticity, context, and regulation: Perspectives from affective neuroscience. *Psychological Bulletin, 126,* 890–909.

Dickey, R. (1992). The management of a case of treatment-resistant paraphilia with a long-acting LHRH agonist. *Canadian Journal of Psychiatry, 37,* 567–569.

Elliot, M. L., & Biever, L. S. (1996). Head injury and sexual dysfunction. *Brain Injury, 10,* 703–717.

Federoff, J. P. (1993). Serotonergic drug treatment of deviant sexual interests. *Annals of Sex Research, 6,* 105–121.

Feiger, A., Kiev, A. Shrivastava, R. K., Wisselink, P. G., & Wilcox, C. S. (1996). Nefazodone versus sertraline in outpatients with major depression: Focus on efficacy, tolerability, and effects on sexual function and satisfaction. *Journal of Clinical Psychiatry, 57*(Suppl. 2), 53–62.

Frohman, E. M., Frohman, T., & Moreault, A. M. (2002). Acquired sexual paraphilia in patients with multiple sclerosis. *Archives of Neurology, 59,* 1006–1010.

Gessa, G. L., & Tagliamonte, A. (1974). Possible role of brain serotonin and dopamine in controlling male sexual behavior. *Advances in Biochemical Psychopharmacology, 11*(0), 217–228.

Goscinski, I., Kwiatkowski, S., Polak, J., Orlowiejska, M., & Partyk, A. (1997). The Kluver-Bucy syndrome. *Journal of Neurological Sciences, 41,* 269–272.

Grant, J. E., & Kim, S. W. (2002). An open label study of naltrexone in the treatment of kleptomania. *Journal of Clinical Psychiatry, 63,* 349–56.

Huws, R., Shubsachs, A. P., & Taylor, P. J. (1991). Hypersexuality, fetishism and multiple sclerosis. *British Journal of Psychiatry, 158,* 280–281.

Kafka, M. P. (1991). Successful antidepressant treatment of nonparaphilic sexual addictions and paraphilias in men. *Journal of Clinical Psychiatry, 52*(2), 60–65.

Kafka, M. P. (1994). Sertraline pharmacotherapy for paraphilias and paraphilia-related disorders: An open trial. *Annals of Clinical Psychiatry, 6,* 189–195.

Kafka, M. P., & Coleman, E. (1991). Serotonin and paraphilias: The convergence of mood, impulse and compulsive disorders [Letter]. *Journal of Clinical Psychopharmacology, 11,* 223–224.

Kafka, M. P., & Prentky, R. (1992). Fluoxetine treatment of nonparaphilic sexual addictions and paraphilias in men. *Journal of Clinical Psychiatry, 53*(10), 351–358.

Khantzian, E. J., Mack, J. E., & Schatzberg, A. F. (1974). Heroin use as an attempt to cope: Clinical observations. *American Journal of Psychiatry, 131,* 160–164.

Kim, S. W., & Grant, J. E. (2001). An open naltrexone treatment study in pathological gambling disorder. *International Journal of Clinical Psychopharmacology, 16,* 285–289.

Kim, S. W., Grant, J. E., Adson, D., & Shin, Y. C. (2001). Double-blind naltrexone and placebo comparison study in the treatment of pathological gambling. *Biological Psychiatry, 49,* 914–921.

Kluver, H., & Bucy, P. C. (1939). Preliminary analysis of functions of the temporal lobes in monkeys. *Archives of Neurological Psychiatry, 42,* 979–1000.

Kolarsky, A., Madlafousek, J., & Novotna, V. (1978). Stimuli eliciting sexual arousal in males who offend adult women: An experimental study. *Archives of Sexual Behavior, 7*(2), 79–87.

Kolb, L. C., Burris, B. C., & Griffiths, S. (1984). Propanolol and clonidine in the treatment of the chronic post traumatic stress disorders of war. In B. A. van der Kolk (Ed.), *Post traumatic stress disorder: Psychological and biological sequelae* (pp. 97–107). Washington, DC: American Psychiatric Press.

Kolb, L. C., & Mutalipassi, L. R. (1982). The conditioned emotional response: A subclass of the chronic and delayed post-traumatic stress disorders. *Psychiatric Annals, 12,* 979–987.

Lane, R. M. (1997). A critical review of selective serotonin reuptake inhibitor-related sexual dysfunction: Incidence, possible aetiology and implications for management. *Journal of Psychopharmacology, 11,* 72–82.

Langevin, R. (1990). Sexual anomalies and the brain. In W. L. Marshall, D. R. Laws, & H. E. Barbaree (Eds.), *Handbook of sexual assault* (pp. 103–113). New York: Plenum.

LeDoux, J. (1998). Fear and the brain: Where have we been, and where are we going? *Biological Psychiatry, 44,* 1229–1238.

Marrazzi, M. A., Bacon, J. P., Kinzie, J., & Luby, E. D. (1995). Naltrexone use in the treatment of anorexia nervosa and bulimia nervosa. *International Clinical Psychopharmacology, 10,* 163–172.

Marson, L., & McKenna, K. E. (1992). A role for 5-hydroxytryptamine in descending inhibition of spinal sexual reflexes. *Experimental Brain Research, 88,* 313–320.

McIntosh, T. K., Vallano, M. L., & Barfield, R. J. (1980). Effects of morphine, B-endorphin and naloxone on catecholamine levels and sexual behavior in the male rat. *Pharmacology, Biochemistry and Behavior, 13,* 435–441.

Mendelson, J. H., & Mello, N. K. (1974). Alcohol, aggression and androgens. In S. H. Frazier (Ed.), *Aggression* (Vol. 52, pp. 225–247). Baltimore: Williams & Wilkins.

Mendez, M. F., Chow, T., Ringman, J., Twitchell, G., & Hinkin, C. H. (2000). Pedophilia and temporal lobe disturbances. *Journal of Neuropsychiatry and Clinical Neurosciences, 12*(1), 71–76.

Mesulam, M. M. (1981). Dissociative states with abnormal temporal lobe EEG: Multiple personality and the illusion of possession. *Archives of Neurology, 38,* 176–181.

Miller, B., Cummings, J. L., McIntyre, H., Ebers, G., & Grode, M. (1986). Hypersexuality or altered sexual preference following brain injury. *Journal of Neurology, Neurosurgery, and Psychiatry, 49,* 867–873.

Modell, J. G., Katholi, C. R., Modell, J. D., & DePalma, R. L. (1997) Comparative sexual side effects of bupropion, fluoxetine, paroxetine, and sertraline. *Clinical Pharmacology and Therapeutics, 61,* 476–487.

Money, J. (1968). Discussion on hormonal inhibition of libido in male sex offenders. In R. P. Michael (Ed.), *Endocrinology and human behaviour* (p. 169). London: Oxford University Press.

Money, J. (1970). Use of androgen-depleting hormone in the treatment of male sex offenders. *Journal of Sex Research, 6,* 165–172.

Money, J. (1980) *Love and love sickness: The science of sex, gender differences, and pair bonding.* Baltimore: Johns Hopkins University Press.

Money, J. (1990). Forensic sexology: Paraphilic serial rape (biastophilia) and lust murder (erotophonophilia) *American Journal of Psychotherapy, 44*(1), 26–36.

Money, J., & Lamarz, M. (1990). *Vandalized lovemaps.* Buffalo, NY: Prometheus Press.

Monga, T. N., Monga, M., Raina, M. S., & Hardjasudarma, M. (1986). Hypersexuality in stroke. *Archives of Physical Medicine and Rehabilitation, 67,* 415–417.

Murphy, M. R. (1981). Methadone reduces sexual performance and sexual motivation in male Syrian golden hamster. *Pharmacology, Biochemistry and Behavior, 14,* 561–567.

Nakada, T., Lee, H., Kwee, I. L., & Lerner, A. M. (1984). Epileptic Kluver-Bucy syndrome: Case report. *Journal of Clinical Psychiatry, 45*(2), 87–88.

Ortego N., Miller, B. L., Itabashi, H., & Cummings, J. L. (1993). Altered sexual behavior with multiple sclerosis: A case report. *Neuropsychiatry, Neuropsychology, & Behavioral Neurology, 6,* 260–264.

Pearce, J., & Miller, E. (1973). *Clinical aspects of dementia.* London: Ballilliere, Tyndall.

Rada, R., Kellner, R., & Winslow, W. (1976). Plasma testosterone and aggressive behavior. *Psychosomatics, 17,* 138–142.

Rada, R. T., Laws, D. R., & Kellner, R. (1976). Plasma testosterone levels in the rapist. *Psychosomatic Medicine, 38,* 257–268.

Raymond, N. C., Coleman, E., Benefield, C., & Miner, M. (2003). Psychiatric comorbidity and compulsive/impulsive traits in compulsive sexual behavior. *Comprehensive Psychiatry, 44*(5), 370–380.

Raymond, N. C., Coleman, E., Ohlerking, F., Christenson, G. A., & Miner, M. (1999). Psychiatric comorbidity in pedophilic sex offenders. *American Journal of Psychiatry, 156,* 786–788.

Raymond, N. C., Grant, J. E., Kim, S. M., & Coleman E. (2002). Treatment of compulsive sexual behavior with naltrexone and serotonin reuptake inhibitors: Two case studies. *International Clinical Psychopharmacology, 17,* 201–205.

Raymond, N. C., Robinson, B. E., Kraft, C., Rittberg, B., & Coleman, E. (2002). Treatment of pedophilia with leuprolide acetate: A case study. *Journal of Psychology and Human Sexuality, 13*(3/4), 79–88.

Rothbaum, B. O., & Foa, E. B. (1996). Cognitive-behavioral therapy for posttraumatic stress disorder. In B. A. van der Kolk, A. C. McFarlane, & L. Weisaeth (Eds.), *Traumatic stress: The effects of overwhelming experience on mind, body and society* (pp. 491–509). New York: Guilford Press.

Sonne, S., Rubey, R., & Brady, K. (1996). Naltrexone treatment of self-injurious thoughts and behaviors. *Journal of Nervous and Mental Disorders, 184,* 192–195.

Stein, D. J., Hollander, E., Anthony, D. T., Schneier, F. R., Fallon, B. A., Liebowitz, M. R., & Klein, D. F. (1992). Serotonergic medications for sexual obsessions, sexual addictions and paraphilias. *Journal of Clinical Psychiatry, 53,* 267–271.

Steinbusch, H. (1981) Distribution of serotonin-immunoreactivity in the central nervous system of the rat-cell bodies and terminals. *Neuroscience, 6,* 557–618.

Stoller, R. J. (1975). *Perversion: The erotic form of hatred.* New York: Pantheon.

Thibaut, F., Cordier, B., & Kuhn, J. M. (1993). Effect of a long-lasting gonadotrophin hormone-releasing hormone agonist in six cases of severe male paraphilia. *Acta Psychiatria Scandanavia, 87,* 445–450.

Thibaut, F., Cordier, B., & Kuhn, J. M. (1996). Gonadotrophin hormone releasing hormone agonist in cases of severe paraphilia: A lifetime treatment? *Psychoneuroendocrinology, 21,* 411–419.

van der Kolk, B. A. (1983). Psychopharmacological issues in posttraumatic stress disorder. *Hospital and Community Psychiatry, 34,* 683–691.

van der Kolk, B. A. (Ed.). (1984). *Post traumatic stress disorder: Psychological and biological sequelae.* Washington, DC: American Psychiatric Press.

van der Kolk, B. A. (1988). The trauma spectrum: The interaction of biological and social events in the genesis of the trauma response. *Journal of Traumatic Stress, 1,* 273–290.

van der Kolk, B. A., Greenberg, M., Boyd, H., & Krystal, J. (1985). Inescapable shock, neurotransmitters, and addiction to trauma: Toward a psychobiology of post traumatic stress. *Biological Psychiatry, 20,* 314–325.

van der Kolk, B. A., McFarlane, A. C., & Weisaeth, L. (Eds.). (1996). *Traumatic stress: The effects of overwhelming experience on mind, body and society.* New York: Guilford Press.

Verebey, K., Volavka, J., & Clouet, D. (1978). Endorphins in psychiatry. *Archives of General Psychiatry, 35,* 877–888.

Volpicelli, J. R., Alterman, A. I., Hayashida, M., & O'Brien, C. P. (1992). Naltrexone in the treatment of alcohol dependence. *Archives of General Psychiatry, 49,* 876–880.

Yehuda, R. (1998). Psychoneuroendocrinology of post-traumatic stress disorder. *Psychiatric Clinics of North America, 21,* 359–379.

7

SYNTHESIS: OVERARCHING THEMES AND FUTURE DIRECTIONS FOR RESEARCH

JANET SHIBLEY HYDE AND BEVERLY WHIPPLE

In this volume, readers have found a dazzling array of results from the very cutting edge of research on the biological substrates of human sexuality. A wide range of behaviors and identities have been considered, including erection, orgasm, compulsive sexual behavior, homosexuality, transsexualism, and orgasm in women with spinal cord injury. The array of biological substrates is equally broad and includes genes, neurotransmitters, hormones, brain regions, the spinal cord, and peripheral nerves. In this final chapter, we synthesize this material, first considering overarching themes across the chapters, and then considering future directions for research.

OVERARCHING THEMES

Several overarching themes emerged across the chapters of this volume: Biological substrates are developmental; sexuality is complex and multifaceted; certain regions of the brain have been identified as key for sexuality, and the spinal cord is crucial as well; and the biological substrates of human sexuality involve multiple interlocking systems.

Biological Substrates Are Developmental

Traditional models regard biological influences as fixed and what people are born with. Modern neuroscience research tells a very different tale. For example, in both rats and humans there are differences between males and females in the sexually dimorphic nucleus of the preoptic area (SDN-POA) in the hypothalamus. However, in humans this difference appears only after the 4th year of age (Swaab & Hofman, 1988)—that is, it is not present from birth! As a second and more well-known example, sex hormone levels rise dramatically at puberty. Biological phenomena may emerge or disappear at various times across the life span.

Sexuality Is Complex and Multifaceted

Behavioral research on human sexuality provides abundant evidence of its complexity. In humans, for example, there can be marked inconsistencies between the individual's sexual identity (heterosexual, homosexual, or bisexual) and the person's sexual orientation defined behaviorally (Lever, Kanouse, Rogers, Carson, & Hertz, 1992). Some women, for example, think of themselves as lesbians but occasionally sleep with men. Other women think of themselves as bisexual yet have had sex only with men. Some men think of themselves as heterosexual yet occasionally have sex with men.

As another example, if we consider male-to-female transsexuals, they fall into two subcategories: androphilic and gynephilic (Blanchard, Dickey, & Jones, 1995). This refers to their sexual orientation following sex change. That is, some male-to-female transsexuals are attracted to men (androphilic), and others are attracted to women (gynephilic). As one can see from questions of sexual orientation and transsexualism, then, sexuality is complex, and the complexity only expands when we consider other aspects of sexuality.

Because human sexuality is so complex, the biological substrates are complex, and simple models will be inadequate. As described in chapter 2, even in the lowly fruitfly *Drosophila*, sexual behavior involves multiple components, including discriminating conspecifics from members of other species, discriminating male from female conspecifics, searching for a mate, courtship (which includes specific behaviors such as tapping, singing, and licking), and copulation. Geneticists have identified at least 14 specific genes that form the substrates for each of these components (Emmons & Lipton, 2003). Human sexuality is vastly more complex, and the biological mechanisms underlying it must be comparably complex.

Key Regions of the Brain

Certain regions of the brain have been implicated repeatedly in research on sexual behavior: the hypothalamus, the bed nucleus of the stria terminalis (BST), the brainstem, and the frontal and temporal lobes.

In the hypothalamus, one crucial region is the SDN-POA, also known as the interstitial nucleus of the anterior hypothalamus (INAH-1). It is termed the sexually dimorphic nucleus because it is approximately twice as large in young adult men compared with women (Swaab, chap. 3, this volume). In animal models, lesions to this region eliminate mounting, intromission, and ejaculation, although erection still occurs (McKenna, 1998). The evidence indicates that there are more androgen receptors in men compared with women in several hypothalamic regions (Swaab, chap. 3, this volume). The paraventricular nucleus (PVN) of the hypothalamus has been implicated in erection in males. Moreover, Komisaruk and Whipple's research (chap. 5, this volume) using functional magnetic resonance imaging (fMRI) indicates activation of the PVN during orgasm in women. Several regions of the limbic system (including amygdala and hippocampus) are activated as well.

The BST is located adjacent to the hypothalamus, septum, and amygdala. One region shows a marked sexual dimorphism. In male-to-female transsexuals, this region is female-sized, suggesting that it may play a role in core gender identity (Swaab, chap. 3, this volume).

As McKenna (chap. 4, this volume) discusses, the brainstem (medulla and pons) is involved in inhibition of spinal sexual reflexes, which are discussed in the next section. Many of these brainstem neurons are serotonergic and may be the mechanisms by which the selective serotonin reuptake inhibitor (SSRI) antidepressants create sexual dysfunctions.

The frontal lobes of the brain are responsible for executive functioning; applied to sexuality, this refers to appropriate inhibition of certain responses, including sexual responses. As Coleman (chap. 6, this volume) notes, compulsive sexual behavior may be linked to abnormalities of the frontal lobes that reduce inhibitory control of behavior. The temporal lobe has been implicated in numerous studies, such as those showing hypersexuality in patients following a stroke that produced lesions to the temporal lobe (Coleman, chap. 6, this volume). Whether these effects are due to damage to the temporal cortex or to the underlying hippocampus and amygdala is unclear.

The Spinal Cord

As McKenna (chap. 4, this volume) discusses, the spinal cord is also crucial to sexual behavior. Both sympathetic and parasympathetic systems are involved. Sensory nerves innervate the sexual organs. In spinal reflex mechanisms, these sensory nerves pass to interneurons at various levels of the spinal cord, which in turn activate neurons involved in sexual functions such as erection and orgasm. Individuals with spinal cord injury and associated difficulties with sexual functioning provide tangible evidence of the role of the spinal cord in sexuality.

Despite this extensive evidence that all sexual pathways lead through the spinal cord, Komisaruk and Whipple (chap. 5, this volume) have found remarkable evidence of a second, alternative pathway in women with complete spinal cord injury and in women without spinal cord injury as well. Some women with diagnosed complete spinal cord injury report experiencing orgasms during sexual activity. Komisaruk and Whipple's research indicates a crucial role for the sensory Vagus nerves, compared with the hypogastric, pudendal, and pelvic nerves that have traditionally been regarded as the transmitters of sexual sensations and enter the spinal cord at T10 or below. Komisaruk and Whipple's research participants, with complete spinal cord injury at T10 or above, should not have been able to experience genital sensations, much less orgasm. Anatomical research indicates that the Vagus extends to the uterus and cervix, and its ascending pathway passes completely outside the spinal cord, up through the neck to a region of the medulla called the nucleus of the solitary tract (NTS). Their fMRI studies indicate that cervical self-stimulation in women with complete spinal cord injury above T10 results in activation of the NTS. Their research with noninjured women indicates activation of the PVN of the hypothalamus during orgasm.

Multiple, Interlocking Systems

Modern neuroscience research pays tribute to the principle that genes, the brain, and the endocrine system can no longer be considered separately. Rather, they are interlocking systems that involve interactive processes. Swaab provided an excellent example in chapter 3. The SRY gene on the Y chromosome directs prenatal differentiation of the testes, which then begin manufacturing testosterone abundantly during the prenatal period. This testosterone then acts on the developing hypothalamus, differentiating it in a sexually dimorphic manner. One characteristic of the male-differentiated hypothalamus is that it contains more androgen receptors than the female-differentiated hypothalamus.

As another example, positron emission tomography (PET) and fMRI studies in humans indicate that the PVN of the hypothalamus is activated during orgasm in women (Komisaruk & Whipple, chap. 5, this volume). The PVN secretes oxytocin, creating a surge of oxytocin in the bloodstream following orgasm, which may account for feelings of a desire for bonding following orgasm (Swaab, chap. 3, this volume). This illustrates how peripheral stimulation and sensations are linked to brain activity, hormone release, and psychological responses.

FUTURE DIRECTIONS

The impressive new research reviewed in this volume encourages us to ask where we will or should go in research in the next decade. Here we suggest several of the strong possibilities.

Advances in Research Technology

The rapid advances in research on the biological substrates of sexuality have been made possible because of enormous advances in research techniques and technologies. Perhaps most important are the advances in scanning technology represented by PET and fMRI. They allow monitoring of specific brain regions in alive, awake, behaving humans and place us light years ahead of traditional techniques such as dissection of cadavers. The time and energy of many researchers will be needed over the next decade to fully exploit the possibilities of these technologies for sex research.

Even these advanced technologies have certain limitations. Measurements are not made in a person's or couple's natural setting. Participants' movement is severely restricted. Moreover, only those persons who are comfortable with sexual self-stimulation in a laboratory setting can provide research data currently. This last limitation leads to a highly selective research sample, which in turn leads to questions about the extent to which the results can be generalized to the rest of the population. Technological advances that overcome these limitations will surely help the field advance.

Advances in animal studies of sexual function are being made with new neuroanatomical techniques. One such technique, the c-fos method, takes advantage of the fact that some neurons express the early intermediate gene c-fos following strong activation. The c-fos can be identified by staining after the experiment. This allows experimenters to identify neurons that were activated during the course of some stimulus, activity, or even complex behavior. This method can be combined with traditional neuroanatomical techniques, such as tract tracing and immunochemistry to identify the connections and neurochemistry of the c-fos labeled neurons. Stimulation and lesion studies can further verify the functional role of the labeled neurons. A limitation of this technique is that not all neurons express c-fos following activation. Thus, the method gives an incomplete picture. Furthermore, great care must be taken in experimental design to ensure that the differences between experimental and control animals are due only to a specific stimulus or behavior. Nonetheless, this technique has led to a number of important advances.

Another novel method is the use of transneuronal labeling techniques. This involves the use of viruses that infect neurons. The virus is injected into a specific organ, where it is picked up by nerve terminals and retrogradely transported back to the cell body. The virus replicates and is shed by the neuron, where it is picked up by nerve terminals synapsing on the infected cell. The process then continues to the next step. This allows identification of neurons that contribute to the innervation of a given organ, even a number of synapses removed. The functional role of neurons identified in this way can be examined by lesion and stimulation experiments. This technique has also provided valuable insights into the innervation of pelvic organs and the central nervous system control of sexual function.

The Importance of Accurate Measurement and Diagnosis

Research on biological substrates has often been hampered by flaws in the measurement or diagnosis of the behavioral phenotype. For example, research on the genetics of sexual orientation has tended to treat all male homosexuals as a single category, despite credible evidence of multiple subcategories (Bell & Weinberg, 1978). Among lesbians, there is the distinction between lifetime lesbians (who have been lesbian ever since they became sexual) and adult lesbians (who were heterosexual and became lesbian in midlife; Valanis et al., 2000). It may be that there is strong genetic influence on one type of homosexuality but not on others. Until we are more proficient in measuring and identifying the behavioral phenotype, genetic research will be hampered.

The Importance of Interdisciplinary Collaborations

Further advances in research on the biological substrates of human sexuality will depend on interdisciplinary collaborations among researchers from psychology and neuroscience and clinicians. Practicing clinicians often have the best knowledge of particular patterns of disturbed sexual behavior that are in need of explanation and treatment. In a sense, they are the frontline observers. Their experience in treating sexual problems and paraphilias with drugs can provide valuable research leads, insofar as drug treatment at the cutting edge is often more a matter of discovering "what works" than of administering a well-validated protocol. Research psychologists have important contributions to make in the measurement and diagnosis of behavior, as well as from their own expertise in neuroscience. Neuroscientists help by importing the latest developments in neuroanatomy, neurophysiology, and endocrinology, for their potential applications in the study of sexuality. These future collaborations should be extremely fruitful, and the complexity of human sexuality provides ample demonstration of the need for multidisciplinary approaches.

Addressing Relationship Issues

Understanding the sexual behavior of the individual, and its biological substrates, is an enormous challenge, but one on which considerable progress has been made in the last decade. Human sexual behavior, however, usually occurs in the context of an interaction between two people. An important next task is to understand better the relational aspects of sexuality, their biological bases in the two individuals, and possible biological interactions between individuals, such as pheromonal communication.

An illustrative example and cautionary tale come from the research and hidden assumptions that led up to the release of Viagra. The research focused on developing a drug that would enable men with erectile disorder to have erections, certainly a worthy goal. In the process, the men's female partners were ignored, or it was assumed that they would be delighted with their men's newfound erections (Potts, Gavey, Grace, & Vares, 2003). Almost as soon as Viagra was released, however, cases began to surface in which partners were not delighted, for several reasons: Some had become quite accustomed to the absence of erections and absence of sexual activity and liked it that way; others felt that they were being coerced into sex by overeager men and their overeager erections. In short, Viagra can create relationship problems. Moreover, it cannot cure erectile disorders that are a result of relationship problems.

Researchers are just beginning to study the effects of having the female partner act as an equal partner in treatment decisions (Leiblum, 2002). Data are beginning to suggest that if the female partner is involved in the initial assessment of erectile disorder and if she is involved in the decision process as to what treatment will be acceptable to both, there is better compliance. Future research on the biological substrates of human sexuality must take on the challenge of addressing the relational aspects of sexuality.

CONCLUSION

Research on the biological substrates of human sexuality has advanced in spectacular ways in the past 15 years. At the same time, we have many miles to go. We must find methods that enable us to study integrated models of biological, psychological, relational, and sociocultural influences. And whatever the final outcomes are in research on neural pathways, neurotransmitters, and hormones, we must not lose sight of the plasticity of these biological systems and the sexual behaviors associated with them.

REFERENCES

Bell, A. P., & Weinberg, M. S. (1978). *Homosexualities*. New York: Simon & Schuster.

Blanchard, R., Dickey, R., & Jones, C. L. (1995). Comparison of height and weight in homosexual versus nonhomosexual gender dysphorics. *Archives of Sexual Behavior, 24*, 543–554.

Emmons, S. W., & Lipton, J. (2003). Genetic basis of male sexual behavior. *Journal of Neurobiology, 54*, 93–110.

Leiblum, S. R. (2002). After sildenafil: Bridging the gap between pharmacologic treatment and satisfying sexual relationships. *Journal of Clinical Psychiatry, 63*(Suppl. 5), 17–22.

Lever, J., Kanouse, D. E., Rogers, W. H., Carson S., & Hertz, R. (1992). Behavior patterns and sexual identity of bisexual males. *Journal of Sex Research, 29*, 141–167.

McKenna, K. E. (1998). Central control of penile erection. *International Journal of Impotence Research, 10*, S25–S34.

Potts, A., Gavey, N., Grace, V., & Vares, T. (2003). The downside of Viagra: Women's experiences and concerns. *Sociology of Health & Illness, 25*, 697–719.

Swaab, D. F., & Hofman, M. A. (1998). Sexual differentiation of the human hypothalamus: Ontogeny of the sexually dimorphic nucleus of the preoptic area. *Developmental Brain Research, 44*, 314–318.

Valanis, B. G., Bowen, D. J., Bassford, T., Whitlock, E., Charney, P., & Carter, R. A. (2000). Sexual orientation and health: Comparisons in the Women's Health Initiative sample. *Archives of Family Medicine, 9*, 843–853.

AUTHOR INDEX

Abdelgadir, S. E., 25, 68
Åberg, H., 42, 57
Acherman, A. E., 91
Achermann, J. C., 23, 56, 67
Ackerman, A. E., 45, 56
Ackman, D., 87, 100
Acuña-Castroviejo, D., 67
Adachi, H., 90, 104
Adaikan, P. G., 84, 98
Adkins-Regan, E., 16, 17, 18
Adler, N. T., 77, 91, 99, 110, 142
Adson, D., 162, 166
Agyei, Y., 11, 18
Ahlenius, S., 84, 92, 96
Airaksinen, J., 144
Åkerlund, M., 45, 61
Al-Attia, H. M., 47, 56
Albert, D. J., 29, 57
Aldasoro, M., 69
Alexander, C. J., 79, 105, 109, 144
Alexandre, L., 92, 97
Alheid, G. F., 49, 63
Allard, J., 86, 92, 97
Allen, L. S., 24, 31, 33, 37–39, 41, 49,
    54, 57
Aloi, J. A., 84, 92
Alster, P., 44, 71
Alterman, A. I., 162, 169
Alvarez, C., 48, 65
Amagai, T., 99
American Psychiatric Association, 148,
    164
American Spinal Injury Association, 140
Ames, M. A., 51, 60
Anderson, A. B. M., 45, 66
Anderson, C., 14, 20, 50, 68
Anderson, C. H., 53, 57
Anderson-Hunt, M., 43, 45, 57
Andersson, K.-E., 27, 66, 75, 76, 92
Andersson, P. O., 45, 71
Andry, C., 103
Antelman, S. M., 81, 95
Anthony, D. T., 169
Arai, Y., 83, 100

Argiolas, A., 44, 57, 66, 86, 88–90, 92,
    102
Arletti, R., 89, 103
Arnold, A. P., 77, 94, 99
Arnow, B. A., 29, 57
Ashwell, K. W. S., 37, 64
Asplund, R., 42, 57
Aston-Jones, G., 81, 106
Athanassiou, K., 45, 64
Austin, C. P., 71, 106
Azadzoi, K. M., 85, 92, 103, 104

Backensto, E. M., 114, 143
Backstrom, J. R., 92, 164
Bacon, J. P., 162, 167
Bailey, J. M., 9, 11–14, 16, 18, 20, 23,
    50, 51, 57
Bailey, J. N., 20
Bailey, M., 50, 65, 68
Bakker, J., 54, 57
Baldwin, P., 140
Balesar, R., 33, 37, 43, 64, 65
Balkin, T. J., 121, 140
Balse-Srinivasan, P., 107
Bancila, M., 83, 84, 92, 156, 164
Bancroft, J., 27, 57, 153, 164
Band, L. C., 99
Bandler, R., 81, 92
Banner, L. L., 57
Barfield, R. J., 83, 87, 90, 92, 101, 157,
    167
Bari, M., 87, 105
Barnes, N. M., 84, 85, 92
Barrington, F. J. F., 81, 93
Bassford, T., 178
Baştürk, M., 71
Batch, J. A., 25, 57
Bauer, H. G., 29, 58
Baum, M. J., 27, 64, 67, 81, 83, 93, 137,
    145
Bazzett, T. J., 86, 93, 107
Beach, F. A., 80, 93
Beaudoin, G., 64

Beauregard, M., 64
Becker, W., 123, 145
Bednarek, M. A., 65
Behbehani, M. M., 81, 93
Bell, A. P., 14, 15, 18, 23, 50, 57, 176, 178
Bell, C., 76, 93
Bell, J. J., 66
Benefield, C., 148, 168
Benelli, A., 89, 103
Benishay, D. S., 14, 18
Benoit, G., 62, 79, 93, 97
Benson, D. F., 28, 65
Bentele, K., 71
Berard, E. J. J., 109, 140
Berendsen, H. H., 84, 90, 93, 107
Berger, B., 48, 65
Berglund, H., 43, 69
Berkley, K. J., 77, 93, 110, 115, 140, 142
Berkovitz, G. D., 72
Berlin, F., 150, 164
Bernabé, J., 79, 89, 92, 93, 97, 104
Bernaschek, G., 63
Berthoud, H. R., 114, 141
Bertolini, A., 89, 93, 103
Bertrand, J., 46, 61
Beyer, C., 53, 59, 140–142
Bianca, R., 114, 140–143
Bieglmayer, C., 58, 141
Biever, L. S., 150, 151, 166
Billerbeck, A. E. C., 59
Bindert, A., 61
Bingham, N. C., 59
Binstock, T., 15, 18
Birder, L. A., 78, 93
Birklein, F., 114, 142
Bitran, D., 83, 84, 87, 93, 99, 155, 164
Björklund, A., 83, 85, 86, 93, 100, 105
Björkstrand, E., 44, 71
Black, D. W., 160, 164
Blaicher, A. M., 44, 58, 141
Blaicher, W., 58, 116, 122, 138, 141
Blake, J., 73
Blanchard, R., 15, 18, 172, 178
Blanco, R., 104
Blaustein, J. D., 83, 106, 137, 144
Blessing, W. W., 37, 61
Blinn, K. A., 79, 94
Bloch, G. J., 37, 58
Blok, B. F., 84, 107
Blomqvist, A., 45, 66

Blumer, D., 151, 165
Bocklandt, S., 50, 60
Boer, K., 60
Boer, R. C., 80, 98
Bogaert, A. F., 15, 18
Bolton, N. J., 24, 58
Bonica, J. J., 110, 141
Bonnefond, C., 37, 58
Booth, A. M., 75–77, 96, 103
Bornhovd, K., 138, 141
Bors, E., 79, 94
Bourgouin, P., 64
Bowen, D. J., 178
Bowser, R., 67
Boyd, H., 155, 169
Boyd, J., 69
Braak, E., 36, 37, 58, 69
Braak, H., 31, 36, 37, 58, 69
Brackett, N. L., 81, 94
Bradford, J., 153, 160, 165
Bradley, S. J., 23, 47, 48, 58, 73
Bradley, W. E., 76, 104
Brady, K., 162, 169
Braun, A. R., 140
Bredt, D. S., 76, 90, 94, 104
Breedlove, S. M., 77, 94, 99
Brindley, G. S., 79, 88, 94
Brockhaus, H., 37, 58
Broekkamp, C. L., 84, 93
Bromm, B., 141
Brown, B. D., 57
Brown, K., 62, 97
Brown, R. L., 90, 96
Brown, T., 65
Brown, T. R., 72
Brown-Grant, K., 53, 58
Buchel, C., 141
Bucy, P. C., 83, 99, 151, 167
Buijs, R. M., 53, 71, 86, 94
Burke, D., 51, 60
Burnett, A. L., 76, 91, 94
Burris, B. C., 155, 167
Burton, S., 44, 67, 88, 103
Butler, P. C., 37, 58
Byne, W., 31, 37, 39, 41, 51, 54, 58, 59

Cador, M., 83, 96
Cadore, L. P., 44, 62
Caggiula, A. R., 81, 95
Calas, A., 82, 106

Calligaro, D. O., 97
Campbell, B., 44, 59
Candemir, Z., 71
Canning, D. A., 48, 72
Card, J. P., 78, 94
Carey, M. P., 88, 94
Carlson, R. R., 77, 94
Carlsson, A., 92
Carlsson, C. A., 79, 96, 100
Carmichael, M. S., 44, 45, 59, 88, 94, 116, 122, 138, 141
Carr, M. C., 48, 72
Carro-Juarez, M., 84, 95
Carson, C., 88, 98
Carson, S., 172, 178
Carter, C. S., 44, 59, 89, 95
Carter, L. S., 43, 59
Carter, R. A., 178
Casey, K. L., 138, 141
Cavazzuti, E., 89, 103
Cechetto, D. F., 82, 95
Cerruti, C., 84, 101
Cervantes, M., 87, 98
Cesnik, J., 149, 161, 165
Chambers, K. C., 86, 95
Chan, J. Y. H., 83, 95
Chan, S. H. H., 82, 95
Chang, L. S., 83, 95
Chang, T. S., 94
Chang, T. S. H., 76, 94
Charney, P., 178
Chawla, M. K., 37, 59
Checkley, S. A., 44, 45, 67, 88, 103
Chen, K. K., 82, 95
Chen, K.-K., 44, 59
Chernick, A. B., 47, 58
Cherny, S., 19, 63
Chien, P. Y., 51, 61
Chivers, M. L., 9, 20
Cho, J. K., 144
Choate, J. V. A., 68
Chow, T., 151, 167
Christ, G. J., 104
Christenson, G. A., 160, 168
Chung, S. K., 80, 101
Chung, W. C. J., 24, 59
Clancy, A. N., 81, 98
Clark, J. T., 87, 95
Clark, T. K., 81, 95
Clattenburg, R. E., 53, 59
Clemens, L. G., 45, 56, 82, 88, 91, 95, 107

Clouet, D., 157, 169
Cohen, R. M., 69
Cohen-Kettenis, P. T., 23, 59, 60
Colapinto, J., 47, 59
Cole, T., 109, 141
Cole-Harding, S., 23, 51, 60
Coleman, E., 147–149, 152, 157, 160–162, 165, 166, 168
Collado, P., 53, 59
Collins, J. J., 114, 141
Comarr, A. E., 79, 94
Condra, M., 88, 103, 104
Coolen, L. M., 79, 83, 95, 106, 137, 144
Coolidge, F. L., 23, 46, 59
Copelin, T., 78, 103
Cordier, B., 162, 169
Correa, R. V., 23, 59
Courtney, K., 27, 62
Courtois, F., 62, 97
Courty, E., 86, 95
Courty, P., 86, 95
Cowan, W. M., 82, 104
Criminale, C., 142
Cristiane, L., 71
Cross, B. A., 27, 60, 122, 138, 141
Crowley, T. J., 90, 95, 158, 165
Cueva-Rolon, R., 114, 140–143
Cummings, J. L., 27, 28, 62, 65, 66, 149, 151, 167, 168
Cunningham, S. T., 110, 141
Curran, T., 78, 104
Cushman, P., 90, 95, 158, 165

Dahl, E., 17, 19
Dalle Ore, G., 28, 70
Dalm, E., 81, 98
Damasio, A. R., 150, 165
Damasio, H., 150, 165
Dammann, O., 71
Damsma, G., 137, 144
Davidson, J. M., 44, 59, 82, 87, 94, 95, 116, 141
Davidson, J. R. T., 160, 165
Davidson, R. J., 156, 166
Davies, H. R., 57
Davis, B. L., 77, 87, 95
Davis, P. G., 91
Dawood, K., 18, 57
Dawson, T. M., 104
Deecher, D. C., 67

De Feo, V. J., 77, *94*
de Groat, W. C., 75, 78, 79, 81, 84, 85, 93, 96, *105*
De Jonge, F. H., 36, *60*
DeLamater, J. D., 11, *19*
Dellovade, T. L., 49, *66*
De Muinck Keizer-Schram, S. M. P. F., *69*
Dennerstein, L., 43, 45, *57*
De Olmos, J., 49, *63*
DePalma, R. L., 80, *102, 156, 167*
Derdinger, F., *92*
Desmond, J. E., *57*
Dessens, A. B., 23, 26, 46, *60*
Devlieger, H., 22, *60*
De Vries, G. J., 24, *59*
De Wall, H., 81, *98*
De Wied, D., *71*
Dewing, P., 26, *60*
De Zegher, F., 22, *60*
Diamond, L. E., 27, 47, *67*
Diamond, M., 12, 20, 23, *60*
Dickey, R., 162, *166, 172, 178*
Dieckmann, G., 29, *60*
Ding, Y. Q., 110, *141*
Dittmann, R. W., 23, 50, *60*
Dixen, J., 44, 59, 116, *141*
Dixon, J., *94*
Doherty, P. C., 84, *98*
Domenech, C., *69*
Domenice, S., *59*
Domer, F. R., 90, *96*
Dominguez, J., *101*
Donker, P. J., 76, *107*
Dörner, G., 39, *60*
Dorr, R., 90, *107*
Downie, J. W., 78, *96*
Drop, M. D., *69*
Droupy, S., *97*
Du, H.-J., 80, 85, *96*
Du, J., *99*
Dun, N. J., 91, *96*
Dun, S. L., *96*
Dündar, M., *71*
Dunne, M. P., 12, *18*
DuPree, M. G., 50, *60*
Durif, F., 86, *95*
Dwyer, S. M., 149, *165*

Earle, D., 27, *67*
Eaton, R. C., 93, 99, *107*

Ebers, G., 14, 20, 27, 66, 68, 149, *167*
Eckersell, C., 37, *58*
Edgar, M. A., *58*
Edwards, D. A., 28, 60, 81, 94, *98*
Ehlhart, T., 54, *70*
Ehrhardt, A. A., 22, 23, 50, 60, *66*
Eippers, B. A., 89, *100*
Elliot, M. L., 150, 151, *166*
Ellis, L., 16, 19, 23, 51, *60*
Emmons, S. W., 16, 19, 172, *178*
Endert, E., 36, 60, *71*
Endo, A., 26, *72*
Enquist, L. W., *94*
Erdman, W. J., II., 79, *104*
Erhardt, A. A., 47, *67*
Ericson, H., 45, *66*
Erlandson, B. E., 79, 96, *100*
Ernst, E., 88, *96*
Erskine, M. S., 83, 96, 100, 137, *142, 144, 145*
Escalante, A. L., 84, *96*
Eşel, E., *71*
Espila, A. M., 37, 64, *65*
Etienne, P., 87, *100*
Evans, B. A. J., *57*
Everitt, B. J., 81–83, 86, 90, 93, 96, *99*
Evers, P., *60*
Exton, M. S., 27, *61*

Fabbri, A., 90, *96*
Facchinetti, P., 82, *106*
Fagerstrom, R. M., 46, *61*
Fakhrai, A., 23, 46, *69*
Fall, M., 79, 96, *100*
Fallon, B. A., *169*
Farrer, L. A., 18, *57*
Faught, R. E., 114, *143*
Federoff, J. P., 157, 160, *166*
Fedirchuk, B., 78, *96*
Feiger, A., 161, *166*
Feldman, J. F., *60*
Felten, D. L., 110, 127, 134, 138, *142*
Ferguson, J. K. W., 138, *142*
Fernández-Guasti, A., 37, 39–42, 49, 61, 65, 84, 87, 96, *104*
Ferrari, W., 89, 93, *97*
Ferrini, M. G., 90, *97*
Fibiger, H. C., 137, *144*
Fillingim, R., 114, *143*
Finkelstein, J. W., *66*

Finn, P. D., 27, *72*
Fisher, C., 25, *67*
Fitzgerald, J. A., *68*
Fleming, S., *73*
Fletcher, T. F., 76, *104*
Fliers, E., 27, 30, 36, 37, 54, *61, 62, 69*
Flumerfelt, D. L., 160, *164*
Foa, E. B., 160, *168*
Fodor, M., 37, 40, 42, *61, 65*
Fong, T. M., *65*
Foreman, M. M., 84–86, *97*
Forest, M. G., 22, 24, 46, *61*
Forsermann, U., *96*
Forsling, M. L., 45, *61, 64*
Frank, G. R., *69*
Frankel, M., 28, *65*
Frankhuijsen-Sierevogel, A. C., *71*
Franzese, A., *96*
Frayne, J., 45, *61*
Freund-Mercier, M.-J., *106*
Frohman, E. M., 27, *61*, 149, *166*
Frohman, T., 149, *166*
Frohman, T. C., 27, *61*
Fuciarelli, K., *107*
Fukai, K., 81, *97*
Fukuda, H., 81, *97*
Fulker, D., *19*
Fulker, D. W., *63*
Fuller, R. W., *97*
Fuller, S. A., *144*
Futreal, P. A., *19*
Futterweit, W., 23, 46, *61*
Fuxe, K., 86, *96*

Gabard, D. L., 17, *19*
Gagnon, J. H., 10, *19*
Gai, W. P., 37, *61*
Gajdusek, D. C., *63*
Ganduglia-Pirovano, M., *140, 142*
Gao, B., 37, *61*
Garcia, M. M., 49, *63*
Garcia-Bazan, M., *143*
Garman, C., *142*
Garrels, L., 46, *61*
Garrick, N. A., 84, *106*
Gaspar, P., 48, *65*
Gaus, S. E., 36, 37, *61*
Gautier, T., 25, *63*
Gavey, N., 177, *178*
Gearhart, J. P., 23, 48, 68, *72*

Gebhard, P. H., 10, *19*
Gebhart, G. F., 114, *144*
Geffen, L. B., 37, *61*
Gehm, B. D., 51, *61*
Georgiadis, J. R., *142*
Gerdes, C., 115, 139, *143, 145*
Gerdes, C. A., 77, *100*, 109, 111–115, 139, *142–144*
Gerhardt, C. C., 84, *97*
Gerlaugh, R. L., 79, *104*
Gessa, G. L., 44, 57, 86, 89, 90, 92, 93, 97, 102, 155, *166*
Getzinger, M. J., 83, *106*, 137, *144*
Gibbs, R. B., 54, *68*
Gilman, S. E., 13, *19*
Ginton, A., 34, 66, 82, *102*
Giovenardi, M., 44, *62*
Giuliano, F., 36, 62, 78, 79, 82, 86, 89, 92, 93, 97, 104, *106*
Gladue, B. A., 39, *62*
Glauche, V., *141*
Glover, G. H., *57*
Glover, T. D., 27, *60*
Gnessi, L., *96*
Goekoop, J. G., *71*
Goetz, C. G., *106*
Gold, M. S., 87, *102*
Goldstein, A., 85, 90, *97*
Goldstein, I., 88, 92, 97, 98, *103*
Goldstein, J. A., 90, *98*
Golombok, S., 23, 51, *62*
Gomez, L. E., *140–142*
Gomora, P., *142*
Gonzales-Cadavid, N. F., 90, *97*
Gonzales-Mariscal, G., *142*
Goodwin, F. K., 87, *98*
Gooren, L. J. G., 22, 23, 27, 31, 35, 49, 50, 55, 59, 62, 65, *70*
Görcs, T., 48, *72*
Gordon, J. H., 30, *62*
Gorman, D. G., 28, *62*
Gorski, R. A., 24, 30, 31, 33, 37, 54, 57, 62, *69*
Gorzalka, B. B., 84, 90, 98, *103*
Goscinski, I., 151, *166*
Goto, Y., 82, *107*
Gottfries, C. G., 33, *62*
Gouras, G. K., 49, *62*
Gower, A. J., 90, *93*
Goy, R. W., 82, *105*
Grace, V., 177, *178*

Grant, J. E., 162, *166*, *168*
Gratzer, T., 160, 161, *165*
Gréco, B., 81, *98*
Green, R., 23, 39, 46, 51, *62*
Greenberg, M., 155, *169*
Greenleaf, W., 59, 94, *141*
Griffin, J. E., 25, *72*
Griffiths, D., 81, *98*
Griffiths, S., 155, *167*
Grimes, S., *143*
Grode, M., 27, 66, 149, *167*
Gross, J., 138, *144*
Gruber, D., 58, *141*
Gruen, R. S., *66*
Grumbach, M. M., 25, *67*
Guan, X., 65, *106*
Guevara, M. A., 87, *98*
Guevara-Guzman, R., *143*
Güldner, F.-H., 53, *62*
Gulyas, B., 43, *69*
Gutierrez, G. M., 37, *59*
Gwinup, G., 88, *98*

Hackett, J., *107*
Hadley, M. E., 90, *107*
Hall, J. L., 86, *97*
Hamann, S. B., 138, *142*
Hamburger-Bar, R., 86, *98*
Hamer, D., 14, *19*
Hamer, D. H., 23, 50, 60, 62, *63*
Hampson, J. G., 47, *67*
Hampson, J. L., 47, *67*
Han, A G., *107*
Hancock, M. B., 76, *98*
Handwerker, H. O., 114, *142*
Hanley, D. F., *94*
Hanrahan, S. B., 137, *142*
Hansen, J., *107*
Hansen, S., 84, *105*
Hanson, L. A., 84, *98*
Hansteen, R. W., 90, *97*
Harada, N., 22, *69*
Haraguchi, N., *70*
Hardjasudarma, M., 151, *168*
Harkness, B., *143*
Harlan, R. E., 49, *63*
Harris, C., *104*
Harsch, H. H., *144*
Harshman, K., *19*
Hart, B. L., 82, 98, *105*

Hartmann, U., *61*
Hasson, H. M., 76, *98*
Hayashida, M., 162, *169*
Hayes, E. S., 84, *98*
Haynes, P. J., 45, *66*
Heath, R. G., 28, *63*
Heaton, J. P., 87, *98*
Hedlund, P., 27, *66*
Heeb, M. M., 137, *144*
Heimann, H., 29, *63*
Heimer, L., 49, *63*
Helke, C. J., 84, *106*
Helleman, R. E., 39, *62*
Hendricks, S. E., 80, *107*, *108*
Hengstschläger, M., 23, *63*
Hepner, B. C., 84, *104*
Herbert, J., 90, *99*
Herman, R. A., 138, *142*
Hernandez-Gonzales, M., 87, *98*
Hertz, R., 172, *178*
Heston, L., 11, *19*
Higuchi, T., *63*
Hillegaart V., 84, *96*
Hines, M., 24, 31, *57*
Hinkin, C. H., 151, *167*
Hiort, O., 21, *63*
Hjorth, S., *92*
Hoffman, N. W., 27, 53, *72*
Hoffmann, R. G., *144*
Hofman, M. A., 22, 24, 27, 30, 31, 33–
    36, 38, 39, 42, 49–51, 54, 55,
    63–65, 70, 72, 172, *178*
Höglund, U., 45, *66*
Hokfelt, F. T., 86, *96*
Hollander, E., *169*
Holman, S. D., 53, 59, *63*
Holmes, G. M., *99*
Holstege, G., 80, 81, 84, 98, *107*
Holstege, J. C., 80, *98*
Honda, C. N., 78, 82, *99*
Honda, K., 43, 63, *107*
Honnebier, M. B. O. M., 51, *70*
Hood, J., *73*
Horita, H., 90, *104*
Horvath, S., 26, 53, *60*
Horvath, T. L., *71*
Hostege, G., 137, *142*
Hotta, H., 110, *140*
Howard, A. D., 65, 71, *106*
Howard, G., 38, *63*
Hruby, V., 90, *107*

Hu, N., 14, *19*, 50, *63*, *65*
Hu, S., 14, *19*, 23, 50, *63*, *65*
Hüber, J. C., *58*, *63*, *141*
Hubscher, C. H., 80, 99, 115, *142*
Hughes, A. M., 90, 99
Hughes, H. E., 48, *73*
Hughes, I. A., 57
Hull, E. M., 83–87, *93*, 99, *101*, *103*,
        107, 155, *164*
Hulse, C., *142*
Humbert, R., *59*, *94*, *141*
Hutchinson, J., 77, 99
Hutchison, J., 110, *142*
Hutchison, J. B., 53, *59*, *63*
Huws, R., 151, *166*
Hwang, P. M., 90, *94*
Hyde, J. F., 38, *63*
Hyde, J. S., 11, *19*

Iadarola, M. J., 78, *93*
Iannucci, U., 89, 90, *102*
Imanishi, J., 99
Imperato-McGinley, J., 23, 25, *63*
Insel, R. T., *144*
Insel, T. R., 43, *64*, *72*, 84, *92*
Ishunina, T. A., 33, 42, 43, *64*
Islam, A., 88, 98
Itabashi, H., 151, *168*
Iuvone, P. M., 81, *94*

Jackson, D. C., 156, *166*
Jacoby, K., *142*
Jameson, J. L., 23, 51, 56, *61*, *67*
Jänig, W., 76, 99
Jannini, E. A., 96
Jardin, A., *62*, *97*
Jeffries, K., *140*
Jenck, F., 84, *93*
Johansson, C., 45, *66*
Johnson, B. T., 88, *94*
Johnson, R. D., 80, 99
Johnson, V. E., *143*
Jones, C. L., 172, *178*
Jonk, R. H., 29, *57*
Jonsson, C., 86, *96*
Jordan, C. L., 77, 99
Jörnvall, H., 37, *70*
Joubert, S., *64*
Jozefowicz, R. F., 110, 127, 138, *142*

Ju, G., *141*
Julien, E., 87, 99

Kafka, M. P., 157, 160, *166*
Kahl, U., *67*
Kalin, N. H., 156, *166*
Kallmann, F. J., 11, *19*, *64*
Kalnin, A., *142*, *143*
Kalra, P. S., 87, *95*
Kalra, S. P., *95*
Kalsbeek, A., 54, *65*
Kamphorst, W., *70*
Kanouse, D. E., 10, *19*, 172, *178*
Kappes, M. E., 50, 60
Kappes, M. H., 50, 60
Karakiulakis, G., 53, *65*
Karama, S., 27, *64*
Kartha, K. N. B., 36, *64*
Katholi, C. R., 80, *102*, 156, *167*
Katz, L.S., 16, *20*
Kawano, H., *69*
Kehrberg, L. L., 160, *164*
Kellner, R., 153, *168*
Kemether, E., *58*
Kendler, K. S., 13, *19*
Kessler, R. C., 13, *19*
Kettl, P., 109, *142*
Keverne, E. B., 23, 46, *62*
Keyes, J. W., Jr., *143*
Khantzian, E. J., 157, *166*
Kidd, G. J., *104*
Kiely, M. E., 87, *100*
Kiev, A., 161, *166*
Kim, S. W., 162, *166*, *168*
Kindon, H. A., 27, 36, *64*
King, M. N., 12, *19*
King, V. L., *65*
Kinsey, A. C., 10, 15, *19*
Kinzie, J., 162, *167*
Kirchner, A., 114, *142*
Kirkpatrick, M., 51, *64*
Kissler, S., 123, *145*
Klassen, P., 15, *18*
Klawans, H. L., *106*
Klein, D. F., *169*
Kleopoulos, S. P., 54, *68*
Kluver, H., 99, 151, *167*
Knogler, W., *58*, *141*
Kockott, G., *61*
Kodman-Jones, C., 48, *72*

Kohlert, J. G., 37, 58
Koivisto, M., 24, 58
Kojima, M., 81, 99
Kolarsky, A., 151, 167
Kolb, L. C., 154, 155, 167
Koliatsos, V. E., 49, 62
Kollar, E. J., 77, 99
Komisaruk, B., 77, 91, 99, 100, 107,
        109–116, 119, 124, 137–140,
        141–145
Kondo, Y., 83, 100
Koppe, J. G., 60
Kordower, J. H., 67
Kostoglou-Athanassiou, I., 45, 64
Koutcherov, Y., 37, 64
Kovenock, J., 24, 71
Kow, L. M., 110, 143
Kow, L.-M., 77, 100
Kraan, E., 42, 65
Kraft, C., 162, 168
Krane, R. J., 85, 92, 103
Kristal, M. B., 110, 144
Krüger, T., 61
Kruijver, F. P., 33, 37, 39, 40–42, 43, 49,
        50, 55, 61, 64, 65
Krystal, J., 155, 169
Kuhn, J. M., 162, 169
Kuikka, J., 144
Kupila, J, 144
Kurohata, T., 90, 104
Kuru, M., 81, 100
Kuwabara, Y., 44, 70
Kuypers, H. G. J. M., 78, 80, 88, 98,
        100, 106
Kwee, I. L., 151, 168
Kwiatkowski, S., 151, 166

Lahr, G., 26, 66
Lake Polan, M., 57
Lakhdar-Ghazal, N., 54, 65
Lal, S., 87, 100
Lamarz, M., 151, 168
Landén, M., 46, 65
Lane, R. M., 80, 84, 85, 100, 156, 167
Lange, G. M., 45, 56, 91
Langevin, R., 150, 167
Langworthy, O. R., 76, 100
Lanta, L., 48, 72
Larkin, K., 28, 68
Larsson, K., 82, 84, 92, 96, 106

Laryea, E., 100
Lasco, M. S., 58
Laumann, E. O., 10, 11, 19
Lavarenne, J., 86, 95
Lavielle, G., 84, 102
Lawrence, B., 66
Laws, D. R., 153, 168
Le Blond, C. B., 53, 65
Lecoq, A., 46, 61
LeDoux, J., 154, 167
Lee, B.-H., 94
Lee, E.-J., 23, 56
Lee, H., 151, 168
Lee, J. W., 78, 100
Leedy, M. G., 82, 98
Leiblum, S. R., 177, 178
Lerner, A. M., 151, 168
Leroux, J.-M., 64
Lesur, A., 49, 65
LeVay, S., 31, 37, 39, 41, 54, 65
Lever, J., 10, 19, 172, 178
Levine, N., 90, 107
Lewis, V. G., 50, 67
Li, J. L., 141
Li, L., 19, 63
Liang, P., 28, 60
Licht, P., 123, 145
Liebowitz, M. R., 169
Lightman, S. L., 44, 45, 67, 88, 103
Lilly, R., 28, 65
Lin, C. E., 114, 141
Lindberg, P., 92
Lindblom, J., 107
Lindstrom, S., 79, 100
Lindvall, O., 85, 86, 93, 100, 105
Ling, N., 89, 100
Lipton, J., 16, 19, 172, 178
List, M. S., 80, 101
Liu, W.-C., 142, 143
Liu, Y.-C., 27, 43, 48, 49, 65
Lizza, E. F., 88, 107, 108
Lluch, S., 69
Loewy, A. D., 80–83, 100, 104
Longueville, F., 89, 97
Lookingland, K. J., 87, 95
Lorenz, J., 138, 141
Lorrain, D. S., 85, 86, 99, 101
Loucks, J. A., 99, 107
Louwerse, A. L., 36, 60, 71
Lowenstein, C. J., 76, 94
Luby, E. D., 162, 167

Lucion, A. B., 44, *62*
Lumley, L. A., *93, 99*
Lundström, B., 46, *65*
Lynn, R. B., *94*

Mack, J. E., 157, *166*
Macke, J. P., 25, *65*
MacLean, P. D., 27, 29, 34, 41, 43, *65*
MacNeil, D. J., 27, *65*
Madlafousek, J., 151, *167*
Magee, T. R., 90, *97*
Magnuson, V., 14, *19*
Magnuson, V. L., 50, *63*
Maguire, M. P., *94*
Mai, J. K., 37, 48, *64, 72*
Mains, R. E., 89, *100*
Maixner, W., 114, *143*
Mallory, B., 79, *105*
Malsbury, C. W., 82, *101*
Mancall, E. L., 28, *66*
Mandrika, I., *107*
Manzanares, J., 87, *95*
Markowski, V. P., *93, 99, 107*
Marlier, L., 84, *101*
Marlowe, W. B., 28, *66*
Marrazzi, M. A., 162, *167*
Marshall, E., 9, *19*
Marson, L., 78, 80–84, 88, *101, 103,*
    106, 156, *167*
Martin, C. E., 10, *19*
Martin, G. F., 80, *101*
Martin, J., 12, *20*
Martin, L. J., 49, *66*
Martin, N. G., 12, *18*
Martin, W. J., *71, 106*
Martínez-León, J. B., *69*
Martin-García, J. A., *67*
Marton, E., *63*
Mash, D. C., *67*
Masters, W. H., *143*
Matsumoto, T., *69*
Matsuura, T., *99*
Matthews, A., 90, *96*
Matuszewich, L., 84–86, *99, 101*
Mayer, A., 26, *66*
Mayer, A. D., 110, *141*
Mayer, B., *103*
McAndrews, J. M., 51, *61*
McBean, A., 147, *165*
McConnel, R. A., 81, *95*

McCurdy, J. R., *103*
McDonald, E., 12, *19*
McDonald, P. G., 81, *92*
McDonald, T. J., 37, *70*
McEwen, B. S., *60*
McFarlane, A. C., 160, *169*
McIntosh, T. K., 83, 87, 90, *101, 167*
McIntyre, H., 27, 66, 149, *167*
McKellar, S., 80, 81, 83, *100*
McKenna, K., 27, 34, 36, 43, 44, 66, 76–
    84, 88, 92, 97, *101–103,* 156,
    164, 167, 173, 178
McLachlan, EM., 76, *99*
McMullen, N. T., 37, 59, *68*
McVary, K. T., 80, *101*
Meade, R. P., *94*
Medina, P., *69*
Meeks, J. J., 23, *56*
Meiners, L. C., *142*
Meisel, R. L., 82, *102*
Melis, M. R., 44, *57, 66,* 86, 88–90, *92,*
    102
Mellenbergh, G. J., *60*
Mello, N. K., 153, *167*
Mendelson, J. H., 153, *167*
Mendez, M. F., 151, *167*
Mendonca, B. B., *59*
Mengod, G., 37, *58*
Merari, A., 34, 66, 82, *102*
Mesulam, M. M., 151, *167*
Meyer, W. J., 46, *66*
Meyer-Bahlburg, H. F. L., 22, 23, 46, 50,
    60, 66, *72*
Meyers, R., 28, *66*
Meyerson, B. J., 45, *66*
Michael, N., *61*
Michael, R. P., 81, *98*
Michael, R. T., 10, *19*
Michaels, S., 10, *19*
Migeon, C. J., *72*
Miki, Y., 9, *19*
Millan, M. J., 84, *102*
Miller, B., 149–151, *167*
Miller, B. L., 27, 28, 66, 151, *168*
Miller, E., 151, *168*
Miller, K. E., 78, *103*
Miller, M., *63*
Miller, M. B., *18, 57*
Miller, N. S., 87, *102*
Mills, R., 37, *58*
Miner, M., 148, 160, *168*

Minoshima, S., 138, *141*
Miselis, R. R., *94*
Mitchell, M. D., 45, 66
Mizuno, M., 44, *70*
Mizusawa, H., 27, 66
Mobbs, C. V., 54, 68
Modell, J. D., 80, *102*, 156, *167*
Modell, J. G., 84, 85, *102*, 156, *167*
Molenaar, J. C., 69
Molin-Carballo, A., 54, 67
Molinoff, P. B., 27, 67
Money, J., 23, 46, 47, 50, 67, *72*, 149,
        151, 157, 162, *168*
Monga, M., 151, *168*
Monga, T. N., *168*
Monsma, F. J., 85, *103*
Montemurro, D. G., 53, 59
Moore, A., 149, *165*
Moore, K. E., 87, 95
Moore, R. Y., 37, *61*
Morales, A., 88, *103, 104*
Morali, G., 87, 98
Moreault, A. M., *61*, 149, *166*
Moretti, C., 96
Morgello, S., 58
Morishima, A., 25, 66, 67
Morris, N. M., 24, 53, *71*
Morris, S., 65
Morrow, T. J., 138, *141*
Moses, J., 99
Mosier, K., *143*
Mosier, K. M., 124, *142*
Muceniece, R., *107*
Mueller, E. A., 84, 92
Mufson, E. J., 49, 67
Müller, D., 29, 67
Multamäki, S., 68
Muñoz-Hoyos, A., 67
Murata, T., 63
Murgia, S., 89, 92
Murphy, D. L., 84, 88, 92, *106*
Murphy, J. J., 79, *104*
Murphy, M. R., 44, 45, 67, *103*, 157, *168*
Murphy, R. L., 18, 57
Mustanski, B. S., 9, *20*, 50, 60
Mutalipassi, L. R., 154, *167*
Mutt, V., 37, *70*

Nadelhaft, I., 76, 77, 79, *101–103*
Nadig, P. W., *104*

Nagura, H., 22, 69
Nair, N. P., *100*
Nakach, N., 27, 68
Nakada, T., 151, *168*
Nakano, Y., 16, *20*
Nargund, R. P., 65, *71*, *106*
Narita, K., 63
Neale, M. C., 11, *18*
Negoro, H., 63, 82, *107*
Negrete, J., *100*
Nelson, D. L., 97
Nescvacil, L., 161, *165*
Ness, T. J., 114, *143*
Neuwalder, H. F., 66
New, M. I., 66
Nicholson, H. D., 45, *61*
Nickolaus, S., 84, *106*
Nievergelt, C., 50, 60
Ninomiya-Alarcon, J. G., *143*
Nobin, A., 85, 93
Nolan, C. L., 138, *142*
Noppen, N. W. A. M., 37, *61*
Norgren, R., 124, *144*
Novotna, V., 151, *167*
Nygren, L. G., 81, *103*

O'Brien, C. P., *169*
Ochoa, B., 47, 67
Ohfu, M., 42, *70*
Ohlerking, F., 160, *168*
Okada, H., 81, 97
Okazaki, M., 42, *70*
Oliver, G., 47, *73*
Oliver, G. D., 58
Olivier, B., 84, *107*
Olson, L., 81, *103*
Ono, H., 42, *70*
Ooms, M. P., 60
Orlowiejska, M., 151, *166*
Oropeza, M. V., 87, 98
Ortega-Villalobos, M., 114, *143*
Ortego, N., 151, *168*
Orthner, H., 29, 67
Ottenweller, J. E., *144*
Over, R., 87, 99
Owen, J., 88, *104*
Owen, J. A., *103*
Ozisik, G., 23, 67
Özkul, Y., *71*

Paans, A. M. J., *142*
Padoin, M. J., 44, *62*
Palacios, J. M., 37, *58*
Palmisano, G., 59, *94*, *141*
Pankiewicz, J., *144*
Papka, R. E., 78, 83, *103*, 114, *141*
Paredes, R. G., 27, 36, 67, *68*
Paredes, R. J., 27, *64*
Park, K., 76, *103*
Parker, K. L., *59*
Parker, R. A., 36, *61*
Parks, C., 51, *57*
Partanen, K., *144*
Partiman, T. S., 54, *69*
Partyk, A., 151, *166*
Pattatucci, A., 19, 50, *63*
Patterson, C., 19, *63*
Patterson, M. N., *57*
Paxinos, G., 37, *64*
Payton, T., 85, *92*
Pearce, J., 151, *168*
Pearson, B., 51, *59*
Peckham, W., 51, *60*
Peglion, J. L., 84, *102*
Pehek, E. A., 86, 99, *103*
Peng, L., 38, *63*
Perkins, A., *68*
Perrin-Monneyron, S., 84, *102*
Peters, H. J. P. W., 83, *95*
Peters, L. C., 110, *144*
Petersen, W. E., 44, *59*
Peterson, R. E., 25, *63*
Petersson, M., 44, *71*
Pévet, P., 54, *65*
Peveto, C. A., 76, *98*
Pfaff, D. W., 54, 68, 77, *100*, 110, *143*
Pfaus, J. G., 54, 68, 90, *103*, 137, *144*
Phoenix, C. H., 86, *95*
Pieribone, V. A., 81, *106*
Pilgrim, C., 26, 27, 66, *68*
Pillard, R. C., 9, 11, 13, 16, 18, *20*, 57, *68*
Pilleri, G., 29, *68*
Pittard, J. T., *104*
Pittler, M. H., 88, *96*
Platt, K. B., 78, *101*, *103*
Ploner, M., 138, *144*
Ploog, D. W., 27, 29, 34, 41, 43, *65*
Poeck, K., 29, *68*
Poggioli, R., 89, *103*
Polak, J., 151, *166*

Pomerantz, S. M., 84, *104*
Pomeroy, W. B., 10, *19*
Pool, C. W., 55, *65*
Post, C., *107*
Potts, A., 177, *178*
Powell, R., 53, *65*
Powers, R. E., 49, *66*
Prendergast, M. A., 80, *107*, *108*
Prentky, R., 160, *166*
Preuss, W., *61*
Price, D. L., 49, *66*
Price, E. O., 16, *20*
Privat, A., 84, *101*, *104*
Probst, A., 37, *58*
Prusis, P., *107*
Punnonen, R., 45, *68*
Puri, P., 78, *103*
Purinton, P. T., 76, *104*
Putnam, S. K., *101*

Qin, K., 25, *67*
Quante, M., *141*
Quigley, A., 22, 25, *68*
Quon, C. Y., 27, *67*

Raadsheer, F. C., *70*
Rada, R., 153, *168*
Rada, R. T., 153, *168*
Raifer, J., 90, *97*
Raina, M. S., 151, *168*
Rainey, W. E., *59*
Raisman, G., 53, *58*
Rajput, A. H., *106*
Ramakrishna, T., 36, *64*
Rampin, O., 78, 79, 82, 89, 92, 93, 97, *104*, *106*, *164*
Ramping, O., *62*
Rance, N. E., 37, 49, 59, 62, *68*
Randich, A., 114, *143*, *144*
Rasmussen, K., *97*
Rattner, W. H., 79, *104*
Ravid, R., *70*
Raymond, N., 147, 148, 160, 161, 162, *165*, *168*
Reid, K., 88, *104*
Reinders, A. A. T. S., *142*
Reiner, W. G., 23, 48, *68*
Reisert, I., 26, 27, 66, *68*

Renter, K., *61*
Repa, C., *63*
Resko, J. A., 25, 28, 68, 69
Rice, G., 14, *20, 23,* 68
Richards, E., 77, *107, 139, 144*
Ridet, J. L., 84, *104*
Rieber, I., 29, 68
Rigter, H., 86, 98
Rinamam, L., *94*
Ringman, J., 151, *167*
Risch, N., 14, *20,* 68
Rittberg, B., 162, *168*
Robbins, A., 77, 93, 110, *140*
Robbins, T. W., 83, 96
Robinson, B. E., 162, *168*
Roch Lecours, A., *64*
Rodriguez-Cabezas, T., *67*
Rodriguez-Manzo, G., 84, 87, 95, *104*
Roeder, F., 29, *67*
Rogers, W. H., 10, *19,* 172, *178*
Rökaeus, Å., 37, *70*
Roland, P., 43, 69
Roos, B. E., 33, *62*
Roppolo, J. R., 78, *93*
Rose, J. D., 81, 82, *104*
Roselli, C. E., 25, 28, 68, 69
Rosen, L. R., 60, 66
Rosen, R., 79, 88, 98, *105, 109, 144*
Rothbaum, B. O., 160, *168*
Rowe, D. W., 137, *144*
Rowley, F., *142*
Roy, R., 51, *64*
Rubey, R., 162, *169*
Russell, D. W., 25, *72*
Rustioni, A., 91, *106*
Rutter, M., 51, *62*

Sachs, B. D., 27, *65,* 82, *102*
Sadeghi, M., 23, 46, 69
Saeger, W., *71*
Saenz de Tejada, I., 85, *104*
Sagar, S. M., 78, *104*
Saito, S., 91, *94, 104*
Salamone, J. D., 27, *65*
Salehi, A., 42, 43, *64*
Salle, B., 46, *61*
Sanders-Bush, E., *92, 164*
Sano, Y., *99*
Sansone, G., 123, *140–143, 144*

Saper, C. B., 36, *61,* 81, 82, 95, *100, 104*
Sar, M., 82, *104*
Sasano, H., 22, 69
Sato, T., 25, 69
Sato, Y., 77, 90, *93, 104,* 110, *140*
Satoh, F., 22, 69
Savic, I., 43, 69
Sawchenko, P. E., 87, 88, *105*
Sayag, N., 27, *72*
Schatzberg, A. F., 157, *166*
Schedlowski, M., *61*
Scheller, F., *61*
Schioth, H. B., 89, 92
Schlosser, S. S., 160, *164*
Schmidt, G., 29, *61,* 69
Schmidt, H. H., 85, 96
Schmidt, R. H., *105*
Schneider, H., 29, 60
Schneider-Jonietz, B., 29, 60
Schneier, F. R., *169*
Schnitzler, A., 138, *144*
Schorsch, E., 29, 69
Schrøder, H. D., 77, 81, 87, *105*
Schultz, C., 31, 69
Schwartz, M., 50, *67*
Scott Young, W., III, 37, 49, 59, *62*
Seckl, J. R., 44, 45, *67,* 88, *103*
Segarra, G., 42, 44, 69
Segraves, K., 87, *105*
Segraves, R. T., 87, *105*
Sekine, K., 69
Shackleton, C., *63*
Shadiack, A. M., 27, *67*
Sharp, F. R., 78, *104*
Sharp, T., 84, 85, 92
Shattuck-Eidens, D., *19*
Shefchyk, S. J., 78, 96
Shen, Y., 85, *103*
Shi, J., *141*
Shi, T., 26, 60
Shields, J., 11, *19*
Shimokochi, M., 81, *105*
Shimshi, M., 66
Shimura, T., 81, *105*
Shin, Y. C., 162, *166*
Shinwari, A., *58*
Shipley, M. T., 81, 92
Shoji, Y., *70*
Shrivastava, R. K., 161, *166*
Shryne, J. E., 24, 30, 31, *57, 62*

Shubsachs, A. P., 151, 166
Sigmundson, K., 23, 47, 60
Sigusch, V., 29, 68
Simerly, R. B., 37, 69, 81, 82, 105
Simpson, A., 90, 95, 158, 165
Simpson, E. R., 25, 67
Singh, R. P., 53, 59
Sipski, M. L., 79, 105, 109, 135, 144
Siroky, M. B., 103
Sjögren, C., 45, 71
Skagerberg, G., 81, 83, 85–87, 100, 105
Skottner, A., 107
Skuse, D. H., 26, 69
Slijper, F. M. E., 23, 46, 69
Slimp, J. C., 82, 105
Slob, A. K., 54, 57
Smith, C., 51, 64
Smith, E. P., 25, 69
Smith, E. R., 87, 95
Snyder, HMcC., 48, 72
Snyder, S. H., 76, 90, 94
Sofuoğlu, S., 71
Sohlström, A., 44, 71
Solano-Flores, L. P., 143
Solomon, A., 57
Song, L., 78, 96
Sonne, S., 162, 169
Southam, A. M., 30, 62
Spano, M. S., 44, 66
Specker, B., 69
Spencer, A., 51, 62
Spirna, P., 87, 105
Stacey, P., 82, 96
Stancampiano, R., 90, 102
Steers, W. D., 75, 79, 81, 84, 85, 96, 105
Stefan, H., 114, 142
Stein, D. J., 160, 169
Stein, E., 17, 20
Stein, E. A., 138, 144
Steinbusch, H., 83, 105, 156, 169
Steinman, J. L., 110, 141, 144
Stellflug, J. N., 28, 68
Stenvert, L. S., 69
Stock, S., 71
Stoller, R. J., 155, 169
Stormshak, F., 28, 68
Strecker, R. E., 36, 61
Strömberg, P., 45, 61
Stuart, C. A., 66
Stumpf, W. E., 82, 104
Sturla, E., 25, 63

Succu, S., 44, 66, 89, 90, 102
Suchindran, C., 22, 71
Surridge, D. H., 104
Suzuki, N., 104
Svensson, L., 84, 92, 105
Swaab, D. F., 22, 24, 26, 27, 29–31, 33–
    40, 42, 43, 49–51, 53–55, 59, 61,
    63–66, 69–72, 172, 178
Swanson, L. W., 37, 69, 81, 82, 87, 88,
    104–106
Swensen, J., 19
Szele, F. G., 84, 106

Tagliamonte, A., 155, 166
Takada, G., 70
Takahashi, H., 69
Takahashi, I., 70
Takahashi, K., 22, 69
Takahashi, T., 22, 70
Takano, K., 42, 70
Takeda, S., 44, 70
Tamir, H., 84, 104
Tanaka, A., 99
Tang, Y., 78, 80, 81, 83, 88, 106
Tanner, C. M., 106
Tapanainen, J., 24, 58
Tate, B. A., 36, 61
Tatemoto, K., 37, 70
Tavtigian, S., 19
Taylor, P. J., 151, 166
Teilhac, J. R., 84, 101
Tepper, M., 77, 107, 139, 144
Terzian, H., 28, 70
Tetel, M. J., 83, 106, 137, 144
Thavundayil, J. X., 87, 100
Thede, L. L., 46, 59
Theodosis, D. T., 89, 106
Thibaut, F., 162, 169
Thiessen, B., 106
Thomas, J. J., 28, 53, 66
Thomas, P. J., 65
Thompson, J. T., 86, 93, 103, 107
Thor, K. B., 84, 106
Thornton, L. M., 13, 19
Tiihonen, J., 137, 144
Tilders, F. J. H., 70
Timmermann, L., 138, 144
Toniolo, M. V., 144
Tranel, D., 150, 165
Trapp, B. D., 104

Travers, S. P., 124, *144*
Treacher, D. F., 45, *64*
Trivedi, S., *18, 57*
Truitt, W. A., 79, *106*
Trujillo, R., *101*
Tseng, L. F., *96*
Tsukamoto, T., 90, *104*
Tucker Halpern, C., 22, *71*
Turan, M., 23, 46, *71*
Turkenburg, J. L., 36, *71*
Turnbull, A. C., 45, *66*
Turner, W. J., *71*
Twitchell, G., 151, *167*
Tzschentke, T., 27, *68*

Uberos-Fernandéz, J., *67*
Udry, J. R., 22, 24, *71*
Uematsu, Y., *69*
Ugolini, G., 78, 100, *106*
Uitti, R. J., 87, *106*
Ulisse, S., *96*
Unmehopa, U. A., 37, 64, *65*
Utsunomiya, H., 42, *70*
Uvnäs-Moberg, K., 44, 45, *71*

Vainio, P., *144*
Valanis, B. G., 176, *178*
Valdueza, J. M., 29, *71*
Vallano, M. L., 90, 101, 157, *167*
Valtschanoff, J. G., 91, *106*
Van, F. H., *66*
Van Bockstaele, E. J., 81, *106*
van Delft, A. M., 84, *93*
Van den Hurk, R., 53, *71*
Van de Poll, N. E., 36, 60, *71*
Vanderah, T. W., *107*
Van der Beek, E. M., 53, *71*
Van der Donk, H. A., 53, *71*
van der Graaf, F. H. C. E., *142*
van der Kolk, B. A., 154, 155, 158–160, 165, *169*
Van der Ploeg, L. H., 27, *71*, 78, 89, *106*
Van der Velde, E. A., *71*
Van Dis, H., 82, *106*
Van Eerdenburg, F J. C. M., 24, *71*
van Heerikhuizen, H., 84, *97*
Van Kemper, G. M. J., *71*
Van Lorden, L., *71*
Van Ophemert, J., 54, *57*

Van Trotsenburg, M., *63*
Vares, T., 177, *178*
Varga, M., *140*
Vargiu, L., 86, 89, 92, *97*
Veening, J. G., 83, 95, 106, 137, *144*
Veldhuis, J. D., 22, *60*
Verebey, K., 157, *169*
Vergé, D., 92, *164*
Veridiano, N. P., 60, *66*
Vermeulen, A., 38, *72*
Vernet, D., 90, *97*
Véronneau-Longueville, F., 82, 88, *106*
Vertes, R. P., 80, *101*
Vihko, R., 24, *58*
Viinamäki, O., *68*
Vila, J. M., *69*
Vilain, E., 26, *60*
Visser, T. J., 37, *61*
Vodusek, D. B., 79, *107*
Volavka, J., 157, *169*
Volpicelli, J. R., 162, *169*
Voogt, J. L., 87, 107, *108*
Vortmeyer, A., *71*

Wagner, C. K., 75, 82, 88, *107*
Wagner, G., 76, *92*
Wakerley, J. B., 122, 138, *141*
Waldinger, M. D., 84, 85, *107*
Wålinder, J., 46, *65*
Walker, A. E., 151, *165*
Walker, P. A., *66*
Wallach, S. J. R., 16, *20*
Wallen, K., 138, *142*
Walsh, M. L., 29, *57*
Walsh, P. C., 76, *107*
Walter, A., 48, *72*
Walter, B., 28, *60*
Waltzer, R., 80, *101*
Wang, D. S., *141*
Wang, Z., 43, *72*
Warburton, V. L., 44, 59, 116, *141*
Ward, O. B., 26, *72*
Ward, R. P., 85, *103*
Ware, J. C., *104*
Warner, R. K., 86, 99, *107*
Watabe, T., 26, *72*
Watanabe, T., *69*
Wayner, M. J., *143*
Webb, A., *66*
Weiller, C., *141*

Weinberg, M. S., 15, *18*, 176, *178*
Weinberg, R. J., 91, *106*
Weisaeth, L., 160, *169*
Weiss, J., 23, 46, *56*
Weiss, R. A., *61*
Wenkstern, D., 137, *144*
Wersinger, S. R., 137, *145*
Wertz, J. M., 84, *104*
Wesensten, N. J., *140*
Wessels, H., 89, 90, *107*
Wessler, G., *96*
Westlin, L., 45, *71*
Weyl, N., 51, *72*
Wheeler, M. J., 45, *64*
Whipple, B., 77, *100*, *107*, 109–116, 119,
        137–139, *141–145*
Whipple, C., 114, 115, *141*
Whitam, F. L., 12, *20*
Whitlock, E., *178*
Wiegant, V. M., 53, *71*
Wiersinga, W. M., 37, *61*
Wikberg, J. E., 89, *107*
Wilcox, C. S., 161, *166*
Wildt, L., 123, *145*
Willerman, L., 51, *57*
Williams, D. M., *57*
Williams, S., 78, *103*
Williams, S. J., *103*
Wilson, C., 81, *92*
Wilson, D. A., *104*
Wilson, J. D., 23, 25, 63, *72*
Winblad, B., 33, *62*
Winslow, J. T., *144*
Winslow, W., 153, *168*

Wisniewski, A. B., 23, 25, *72*
Wisselink, P. G., 161, *166*
Wu, S. Y., *96*

Xu, J. Q., *141*

Yahr, P., 27, *72*
Yamaguchi, H., *94*
Yamamoto, D., 16, *20*
Yamane, M., 81, *97*
Yamanouchi, K., 83, *100*
Yanagimoto, M., 63, 82, *107*
Yang, S. P., 87, *107*, *108*
Yehuda, R., 154, 158, *169*
Yells, D. P., 80, 83, *107*, *108*
Young, L. J., 43, *72*
Young, S. E., 46, *59*
Young, W. S., IIII., 68

Zaborszky, L., *142*
Zarefoss, S., *142*
Zderic, S. A., 23, 48, *72*
Zenchak, J. J., 16, *20*
Zenut, M., 86, *95*
Zhang, L., *97*
Zhou, J. N., 31, 35, 49, 50, 53–55, 65,
        70, *72*
Zimmerman, I., *60*
Zorgniotti, A. W., 88, *107*, *108*
Zucker, K. J., 23, 46–48, 50, 51, 58, *72*,
        73

SUBJECT INDEX

A5 catecholaminergic cell group, 81
Abuse, CSB and, 152, 154, 155
ACTH (adrenocorticotropin), 89
Activating effects of sex hormones, 55
Adoptive sibling studies, 11
Adrenal, 24
Adrenocorticotropin (ACTH), 89
Age-related differences
    in INAH-2, 41
    in SDN-POA cell number, 33
    in SON, 42, 43
Aging, sex difference in pattern of, 38–39
Agyei, Y., 11–12
Alcohol
    prenatal exposure to, 26
    and testosterone, 153
Alpha-1 adrenoceptor antagonists, 87, 88
Alpha-2 adrenoceptor agonists, 87, 88
Alpha-melanocyte-stimulating hormones
    (alpha-MSH), 89
Amphetamine, 87, 157
Amygdala, 83, 154
    aromatase in, 22
    BST connections with, 49
    and CSB, 149
    and erectile dysfunction, 27
    orgasm and activation of, 127, 132,
        133, 137, 138
    and sexual orientation, 28
Analgesia, self-stimulation-induced, 113,
    114, 116
Anal sex, 91
Androgen insensitivity syndrome, 25
Androgen receptors, 25
    in BST, 48, 49
    in hypothalamus, 82, 174
    in midbrain, 81
    in SDN-POA, 37
    sex differences in, 41, 173
Androgens
    and male gender identity, 48
    and sexual differentiation of brain,
        22, 24, 25
Androphilic transsexuality, 172

Anterior cingulate cortex, 138
Anterior commissure, 30–32, 51, 134
Anterior hypothalamus
    during arousal, 121–123
    sex differences in, 31
Antiandrogen treatment, 153–154, 162
Anticonvulsants, prenatal exposure to,
    26
Antidepressants, 85, 160–162, 173
Anxiety, 153
Apomorphine, 85–87
Aromatase, 22
Aromatization theory, 21–22, 24, 25
Australian twin registry study, 12
Autoerotic asphyxia, 161
Autonomic function
    and CSB, 154–155
    and nitric oxide, 91
Axons, 80

Bailey, J. M., 11–14
Bancroft, John, 153
Barrington's nucleus, 81
Basal ganglia
    during arousal, 121, 122
    orgasm and activation of, 137
Bed nucleus of the stria terminalis
        (BST)
    dspm area of, 31, 49
    and erectile functions, 27
    estrogen/androgen receptors in, 48
    lesions of, in rats, 28
    orgasm and activation of, 127, 133,
        137
    principal sectors in, 49
    in rats, 48–49
    sexual-behavior regions of, 173
    and sexual differentiation of brain,
        24
    and transsexuality, 45–50
    in transsexuals, 173
Bell, A. P., 14
Benishay, D. S., 14

Bisexuality
    and diethylstilboestrol exposure of
        mother, 22
    in females, 50
    in lesbians vs. gay men, 15
    in sibling studies, 14
Bladder activity inhibition, 79
Bladder control, 81
Brain
    regions activated during orgasm,
        127–138
    sexual-behavior regions of, 173
    sexual differentiation of. *See* Sexual
        differentiation of brain
    weight of, 24
Brain-imaging studies, 109–140
    animal study evidence for
        hypothesis, 114–115
    brain regions activated during
        orgasm, 137–138
    fMRI method in, 123–136
    genital sensory nerves, functions of,
        110
    implications for practitioners,
        139–140
    methodology, 112–113
    pain measurement in, 111–112
    PET method in, 115–123
    tactile thresholds in, 111–112
    Vagus nerves, 111–112, 135, 137
Brainstem, sexual-behavior regions of,
    173
BST. *See* Bed nucleus of the stria
    terminalis
BSTc. *See* Central nucleus of BST
BST-dspm. *See* Darkly staining
    posteromedial of BST
Bulbocavernosus reflex, 79
Bulbospongiosus muscles, 79

Carbamazapine, 152, 161
Caudate, 138
Cavernous nerve, 76, 79, 80, 84
Central gray matter
    and interneurons, 78, 79
    orgasm and activation of, 137
    sensory mechanisms of, 77, 78
Central nervous system, sexual function
    control in, 75–83
    forebrain, 83

hypothalamus, 82–83
interneurons, 78–79
medulla, 80
midbrain, 81
motor mechanisms, 76–77
pons, 81
sensory mechanisms, 77–78
spinal reflexes, 79–80
Central nucleus of BST (BSTc), 31, 35,
    49–50
Cerebellum, 137, 138
Cervical self-stimulation, spinal cord
    injury and. *See* Brain-imaging
    studies
Cervix, 76
C-fos gene, 78
C-fos method, 175–176
Child abuse, CSB and, 154, 155
Child molesters, 153
1-(3-Chlorophenyl)-piperazine (m-CPP),
    84
Chromosomal abnormalities, 46
Cingulate cortex, 127, 132, 133, 137,
    138
Climax, 84
Clitoral priapism, 85
Clitoris, 76, 79
Cloacal exstrophy, 47
Clonidine, 87
Cocaine, 87, 157
Cognition, sex differences in, 29
Cognitive-genital dissociation, 139. *See
    also* Spinal cord injury
Commissura anterior, 51
Complete androgen insensitivity
    syndrome, 25
Compulsive sexual behavior (CSB),
    147–178
    and autonomic reactivity/
        neurotransmitter dysregulation,
        154–156
    and dopamine dysregulation, 157
    and frontal lobe abnormalities, 150
    neuroanatomical abnormalities with,
        149–152
    neurophysiochemical abnormalities
        with, 152–160
    nonparaphilic, 148
    and norepinephrine dysregulation,
        155, 156
    and opioids, 157–160

paraphilic, 148–149
pharmacological treatment of,
160–163
and serotonin dysregulation,
156–157
and temporal lobe abnormalities,
150–152
and testosterone abnormalities,
152–154
Computer tomography imaging (CT),
150
Concordant twin pairs, 10, 12
Congenital adrenal hyperplasia, 46, 50
Copulatory behavior
and dopamine, 85, 86
mPOA and, 27–28, 34, 36
and nitric oxide, 90
and noradrenaline, 87–88
and opiates, 90
and serotonin, 84
Courtless gene (fruit flies), 16
CSB. *See* Compulsive sexual behavior
CT (computer tomography imaging), 150
CYP19 gene, 50
Cyproterone acetate, 162

D1 receptors, 85–86
D2 receptors, 84–86
Darkly staining posteromedial of BST
(BST-dspm), 30, 31, 49
Defecation, 81
DES. *See* Diethylstilboestrol
Diagnosis of sexual orientations, 176
*Diagnostic and Statistical Manual of Mental
Disorders (DSM-IV-TR)*, 148
Diamond, M., 12
Diencephalic dopaminergic cells, 85
Diethylstilboestrol (DES), 50
Discordant twin pairs, 10, 12
Dopamine
and CSB, 155, 157
and nitric oxide, 90
in PVN, erection function and, 44
and sexual function, 84–87
Dorsal horn, 77, 78
Drosophila, 16, 172
Drugs, prenatal exposure to, 26
*DSM-IV-TR (Diagnostic and Statistical
Manual of Mental Disorders)*, 148
Dunne, M. P., 12

Ejaculation
and dopamine, 85
and opiates, 157
and serotonin, 84, 85
and spinal reflex, 79
and SSRIs, 156
Emotional abuse, CSB and, 152
Endocrine system. *See*
Hypothalamus
Endogenous opiates, 157
Endorphins, 157–159
Enkephalins, 157
Ephebophilia, 29
Erectile functions
and biologically active peptides,
89–90
and brain regions, 173
and dopamine, 86, 87
and hypothalamus/amygdala, 27
and nitric oxide, 90
and noradrenaline, 88
and opiates, 90
and oxytocin, 88–89
and preoptic area, 34, 36
and serotonin, 84
and spinal reflex, 79
and supraoptic/paraventricular
nucleus, 43–44
Erotomania, 147
Estrogen receptors
in amygdala, 28
in BST, 48
in SDN-POA, 37–38
sensory mechanisms of, 77
sex differences in, 42
Estrogens, sexual differentiation of brain
and, 22, 24–25
Ethical issues with genetic technologies,
17
Excitatory pathways, 75–76, 86
Ex copula erections, 80
Exhibitionism, 29

Familial studies of sexual orientation,
13–14
Fear response, 154, 155
Females
brain regions activated during
orgasm in, 127–138
clitoral priapism in, 85

Females, *continued*
    complete androgen insensitivity
        syndrome in, 25
    congenital adrenal hyperplasia in, 46
    copulatory behavior in, 27–28
    and dopamine, 86–87
    motoneurons in, 77
    and noradrenaline, 87–88
    and oxytocin, 88, 89
    oxytocin levels in, 45
    sensory mechanisms in, 77
    and sex differences in hypothalamus.
        See Hypothalamus
    and sex differences in SDN-POA,
        172
    sexual differentiation of brain in, 22,
        24–25
    with spinal cord injury, sexual
        response in. *See* Brain-imaging
        studies
    spinal reflexes in, 79–80
    vasopressin levels in, 45
Ferguson reflex, 138
Ferrets, 36
Fetishes, 150
Fight-or-flight responses, 149
fMRI. *See* Functional magnetic resonance
    imaging
Forebrain, 83
Fornix, 30
Fos gene product, 78, 83
Frontal cortex, 137
Frontal lobe
    and CSB, 149, 150
    sexual-behavior regions of, 173
    and sexuality, 173
Fruitless gene (fruit flies), 16
Fuguelike state, 151, 157
Functional magnetic resonance imaging
    (fMRI), 115, 116, 123–136

GABA (gamma-amino butyric acid), 37
Galanin, 36, 37, 49
Gamma-amino butyric acid (GABA), 37
Gay men
    genetic influences on lesbians vs., 15
    hypothalamus differentiation in, 39
    linkage studies of, 14
    and maternal stress, 51
    pedigree studies of, 14

and prenatal nicotine exposure, 51
SCN differences in, 51–53
SDN-POA in, 39
sibling studies of, 14
subtypes of, 15
twin study concordance rates for, 16
Xq29 gene in, 14
Gender identity. *See also* Transsexuality
    and brain–sex hormone interaction,
        48–49
    BST role in, 173
    factors influencing, 23
    genome scan for, 15
    genomewide scan, 15
    in females, 25
    and gender reassignment, 47–48
    in males, 25
Gender imprinting, 47
Genetics of sexual orientation, 9–18, 172
    animal model research on, 16–17
    ethical issues related to, 17
    gender differences in, 15–16
    non-twin familial studies, 13–14
    and single-gene hypothesis, 14
    twin and adoption studies, 10–13
    and types of homosexualities, 15
Genital sensory pathways. *See also*
        Brain-imaging studies
    division of labor among nerves, 110
    and spinal cord injury, 109, 174
    via Vagus nerves, 111–112, 135, 137
Genital sensory stimulation, 82, 89
"Genitosensory Vagus" hypothesis,
    113–115
Gerbils, 53
Gilman, S. E., 13
Golgi apparatus
    age-related differences in, 43
    sex differences in, 42, 43
Gonadotropins, 29
Grafenberg spot, 112
Gray matter, 76–79
    orgasm and activation of central,
        137
    oxytocin in, 88
    periaqueductal, 81
Gynephilic transsexuality, 172

Hamartoma, 29
Hermaphroditism, 48

Heterosexuality
    and androgen action on brain, 25
    factors influencing, 23
    and lesions in hypothalamus, 29
Hippocampus
    aromatase in, 22
    and CSB, 149, 155, 156
    orgasm and activation of, 135–138
Homosexualities
    factors influencing, 23
    gender differences in, 15–16
    and hypothalmic structures, 50–54
    in lesbians vs. gay men, 15
    and lesions of hypothalamus/
        temporal lobe, 28, 29
    in men. *See* Gay men
    in sibling studies, 14
    twin studies of, 10–11
    types of, 15
    in women. *See* Lesbians
Hormonal cues, 82
Hormone-independent sexual
        differentiation, 26
Hormones. *See* Sex hormones
H-Y antigen, 15
8-Hydroxy-2-(di-n-propylamino) tetralin
        (8-OH DPAT), 84
Hypersexual behavior, 149–151
Hypersexuality, 29, 147
Hypogastric nerves, 79, 110
Hypogonadism, 29
Hypothalamus, 27–55, 82–83
    androgen receptors in, 174
    anterior, 31, 121–123
    aromatase in, 22
    during arousal, 121–123
    BST and transsexuality, 45–50
    and CSB, 157
    development/sexual differentiation of
        SDN-POA, 38–40
    homology of human to rat
        SDN-POA, 36–38
    and homosexuality, 50–54
    and noradrenaline, 87
    orgasm and activation of, 127, 132,
        133, 137–138
    sex hormone receptor distribution
        differences, 41–42
    sexual-behavior regions of, 173
    sexual differentiation of, 24
    and sexual function control, 82–83

    sexually dimorphic nucleus of
        preoptic area, 33–36
    sexually dimorphic structures in,
        30–33
    structural differences in, 39, 41
    supraoptic and paraventricular nu-
        cleus, 42–45
    ventromedial nucleus of, 29

Identity, sexual, 10, 172
Immune system, 15
INAH. *See* Interstitial nucleus of anterior
        hypothalamus
Incertohypothalamic pathway, 85
Inescapable shock, 155
Infundibular nucleus, 31
Infundibulum, 30
Insula, 133
Insular cortex, 127, 132, 133, 137, 138
Interdisciplinary collaborations, 176–
        177
Intermediate nucleus, 37
Intermediolateral cell column, 78
Interneurons, 78–79, 81
Interstitial nucleus of anterior
        hypothalamus (INAH), 30, 31,
        37–39, 41, 51, 54
Ischiocavernosus muscles, 79

James I, King of England, 51

Kallman, F. J., 11
Kendler, K. S., 13
Kessler, R. C., 13
King, M. N., 12
Kinsey, Alfred, 9–10
Kinsey rating scale, 10, 15
Klüver-Bucy syndrome, 28, 151

L3 spinal segment, 79
L4 spinal segment, 79
L6-S1 spinal cord, 88
Lactation, oxytocin and, 44
Lateral ventricle, 30
Lesbians
    and congenital adrenal hyperplasia,
        50

Lesbians, *continued*
    and diethylstilboestrol exposure of
      mother, 22
    genetic influences on gay men vs., 15
    genetic studies of, 11–12
    and prenatal nicotine exposure, 51
    and sex hormones, 50–51
    sibling studies of, 13–14
    subtypes of, 15
    twin study concordance rates for, 16
    undertheorization concerning, 16
Leuprolide acetate, 162
LHRH. *See* Luteinizing hormone-
    releasing hormone
Limbic system
    and CSB, 149
    and opioids, 157
    orgasm and activation of, 137, 173
    sex differences in, 27–28
Linkage analysis, 14
Lithium carbonate, 161
Locus ceruleus, 81, 87
Lordosis response, 45
Lower brainstem, 137
Lumbosacral spinal cord, 77, 78, 80, 85, 88
Luprolide acetate, 162
Luteinizing hormone-releasing hormone
    (LHRH), 49, 53, 157, 162

MacArthur Foundation, 13
Magnetic resonance imaging (MRI), 150.
    See also Functional magnetic
    resonance imaging
Male pseudohermaphroditism, 47
Males
    brain regions activated during
      orgasm in, 137–138
    copulatory behavior in, 27–28
    and dopamine, 86, 87
    erectile functions in, 27
    gender identity in, 25
    heterosexuality development in, 25
    and hypothalamus, 82
    motoneurons in, 77
    and noradrenaline, 87, 88
    and oxytocin, 88–89
    prenatal alcohol exposure in, 26
    priapism in, 85
    and sex differences in hypothalamus.
      See Hypothalamus

and sex differences in SDN-POA, 172
    sexual differentiation of brain in, 22,
      24–26
    spinal reflexes in, 79–80
Martin, J., 12
Mary, Queen of Scots, 51
Maternal stress, 51
MC4 receptors, 89
M-CCP, 85
McDonald, E., 12
MC receptors. *See* Melanocortin
    receptors
Measurement
    accuracy in, 176
    advanced technologies for, 175
    of pain, 111–112
    of sexual orientation, 9–10
    of testosterone, 153
Medial amygdala, 27
Medial-basal hypothalamic dopamine
    neurons, 27
Medial mamillary nucleus (MMN),
    41–42
Medial preoptic area (MPOA)
    and copulatory behavior, 27, 34, 36
    and dopamine, 85–87
    and hypothalamus, 82
    and nitric oxide, 90–91
Medroxyprogesterone acetate, 162
Medulla, 80
Melanocortin agonists, 27
Melanocortin (MC) receptors, 78, 89–90
Melanotan II, 89–90
Mesencephalic reticular formation, 137
5-Methoxytryptophol, 54
Mice, 26
Midbrain, 81
Midbrain reticular formation, 120–122
Midlife Development in the United
    States (MIDUS) survey, 13
MIDUS survey. *See* Midlife Development
    in the United States survey
Minnesota Twin Registry study, 12
MMN. *See* Medial mamillary nucleus
Monkeys
    erection functions in, 27, 29
    hypothalmic structures in, 54
    PVN and erection in, 43
    SDN-POA in, 36–37
Mood stabilizers, 161
Motor mechanisms, 76–77

mPOA. *See* Medial preoptic area
MRI (magnetic resonance imaging), 150
MTII, 27

Naloxon, 44
Naloxone, 90
Naltrexone, 90, 162–163
National Health and Social Life Survey,
    10
Neale, M. C., 11–12
Nefazadone, 161
Neocortex, 137
Nerves, genital sensory, 110, 174
Neuroanatomical abnormalities, 149–152
Neurological diseases, 29
Neurological pharmacological agents,
    83–91
    adrenocorticotropin, 89–90
    biologically active peptides, 89–90
    dopamine, 85–87
    nitric oxide, 90–91
    noradrenaline, 87–88
    opiates, 90
    oxytocin, 88–89
    serotonin, 83–85
Neurophysiochemical abnormalities,
    152–160
    autonomic reactivity/
        neurotransmitter dysregulation,
        154–156
    norepinephrine dysregulation, 155,
        156
    opioids, 157–160
    serotonin/dopamine dysregulation,
        156–157
    testosterone, 152–154
Neurotransmitter dysregulation, 158–159
Neurotransmitters. *See also* Serotonin,
        Dopamine
    CSB and dysregulation of, 155, 156
    metabolism of, 26
Neurotrophin receptors, 43
Nitric oxide, 76, 90–91
Nitric oxide synthase (NOS), 90
NO, 76, 90–91
Nocturnal tumescence, 85
Nonparaphilic CSB, 148
Noradrenaline, 87–88
Norepinephrine, 155–156
NOS (nitric oxide synthase), 90

NPGI (nucleus paragigantocellularis), 80
N-trifluoromethylphenylpiperazine
        (TFMPP), 84
NTS. *See* Nucleus of the solitary tract
Nucleus accumbens, orgasm and
        activation of, 133, 137, 138
Nucleus of the solitary tract (NTS), 115,
        116
    fMRI imaging of, 123–136
    orgasm and activation of, 137
    PET imaging of, 116–120
Nucleus paragigantocellularis (nPGi), 80
Nymphomania, 147

Obesity, 89
Obsessive–compulsive disorder, 159
1A receptor subtype, 84, 85
1B receptor subtype, 84
Opiates, 90, 157
Opioids, 157–160
Optic chiasm, 30, 31
Optic tract, 30
Organum vasculosum lamina terminalis,
        31
Orgasm, 85
    brain regions activated during,
        127–138
    and oxytocin levels, 122–123
    and SSRIs, 156
Orgasms
    and oxytocin levels, 44–45
    in women with spinal cord injury.
        See Brain-imaging studies
Osteopenia, 162
Oxytocin, 82, 86, 88–89
    and human maternal behavior, 44
    and nitric oxide, 90, 91
    orgasm and levels of, 122–123
    orgasm and secretion of, 138, 174
    in PVN and SON, 43–44
    sex differences in, 43–45
Oxytocinergic neurons (of PVN), 34, 36
Oxytocin receptor, 45

P75 neurotrophin receptor (p75^{NTR}), 27
Pain
    brain regions activated by, 138
    threshold of, in brain-imaging
        studies, 111–113

Pair bonding, 16, 89
Paleomammalian cortex, 149
Paraphilic CSB, 148–149
   and brain abnormalities, 150
   and temporal lobe abnormalities,
     150–151
Parasympathetic preganglionic neurons,
     76, 81
Paraventricular nucleus (PVN), 30, 31
   activation of, in women with spinal
     cord injury, 116
   during arousal, 121–123
   and dopamine, 85, 86
   and erectile functions, 27, 34, 36,
     173
   and hypothalamus, 82–83
   and nitric oxide, 90
   orgasm and activation of, 127, 134,
     137, 138, 174
   and orgasm in women, 173
   and oxytocin, 88–89
   sex differences in, 42–45
Parietal cortex, 137
Parkinsonism, 86, 157
Parturition, 81
Pedigree analysis, 14
Pedophilia
   and brain injury, 149, 150
   and lesions of hypothalamus/
     temporal lobe, 28, 29
   and temporal lobe abnormalities,
     150–152
Pelvic function, 91
Pelvic ganglion, 76
Pelvic innervation, 76–78
Pelvic interneurons, 78
Pelvic nerves, 110
Penile ablation, 47
Penis, 76, 78
Peptide systems, 158
Periaqueductal gray, 81
Period gene (fruit flies), 16
PET. See Positron emission tomography
Pharmacological treatment (of CSB),
     160–163
Pharmacology of sexual function. See
     Neurological pharmacological
     agents
Phentolamine, 88
Phenylpiperazine reuptake inhibitor
     effects, 161

Pheromones, 43
Phosphodiesterase Type 5 inhibitors, 88
Physical abuse, CSB and, 152
Phytoestrogens, 51
Pillard, R. C., 11–12
Pineal region tumors, 29
Pituitary, 42
Plasma prolactin concentrations, 27
Pleasure, brain regions activated by, 138
POA. See Preoptic area
Polygenic traits, 14
Polysynaptic reflex, 79
Pons (pontine micturition center), 22,
     81
"Poppers," 91
Positron emission tomography (PET),
     115–123, 150
Postganglionic neurons, 76
Posttraumatic stress disorder (PTSD),
     154–156, 158
Prazosin, 87
Precocious puberty, 29
Preganglionic neurons, 76–77
Pregnancy, induction of, 77
Preoptic area (POA). See also Sexually
     dimorphic nucleus of preoptic
     area
   medial, copulatory behavior and, 27
   sex differences in, 31
   and sexual orientation, 28
Priapism, 85
Pro-opiomelanocortin, 89
Prostate, 76
Proust, Marcel, 51
Pseudohermaphroditism, 47
Pseudopregnancy, 77
Pseudorabies virus, 78, 88
Psychiatric diseases, 29
PT 141, 27, 90
PTSD. See Posttraumatic stress disorder
Pudendal nerve, 77–79, 81
Pudendal nerves, 110
Putamen, 137
PVN. See Paraventricular nucleus
Pyriform cortex, 83

Rapists, 153
Rats
   BST in, 48–49
   erectile functions in, 27

female, genital sensory nerves in, 110
female, orgasm in, 137
neurons of NTS in, 115
oxytocin and sexual behaviors in, 44, 45
partner preference and BST lesions in, 28
preoptic area in, 36
PVN lesions in, 43
SCN differences in, 53, 54
SDN-POA in humans vs., 36–38
sex differences in SDN-POA in, 172
sexual behavior and exposure to drugs, 26
SON of, 43
Vagus nerve pathway in, 114–115
vasopressin and copulatory behavior in, 45
Reflexes, spinal, 79–80, 173
Registry studies of sexual orientation, 12–13
Relationship issues, 177
Reproduction, SCN role in, 53, 54
Research technology, advances in, 175–176
Robinson, Gene, 9
Rodents, 24, 25. See also Mice; Rats
RSD 992, 84

Sacral parasympathetic nucleus, 77, 88
Satyriasis, 147
SCN. See Suprachiasmatic nucleus
SDN. See Sexually dimorphic nucleus
SDN-POA. See Sexually dimorphic nucleus of preoptic area
Seizures, 151
Selective serotonin reuptake inhibitors (SSRIs), 160–162, 173
and medulla/pons, 80
and serotonin, 84, 85, 156–157
Selegiline, 86
Self-medication, 157
Sensory cues, 82
Sensory mechanisms, 77–78
Septum, 149
lesions in, 27
orgasm and activation of, 133
and sexual behavior, 28–29

Serotonin, 83–85
and CSB, 155
CSB and dysregulation of, 156–157
and medulla/pons, 80
Serotonin 5 HT$_2$ receptor subtype, 161
7-point scale of sexual orientation, 10
Sex hormone receptors, 41–42. See also specific hormone receptors
Sex hormones. See also specific hormones
activating effects of, 55
decrease in SDN-POA cells and, 38
interaction of developing brain and, 39
interaction of oxytocin and, 45
metabolism of, 26
puberty and levels of, 172
and sexual differentiation of brain, 22–26
and sexual orientation, 50–51
Sexual addiction, 147
Sexual behavior
compulsive. See Compulsive sexual behavior
and exposure to drugs, 26
and galanin in medial preoptic nucleus, 37
hypothalamus and adjoining structures, role of, 28–29
relationship issues in, 177
SCN role in, 53, 54
sexual identity vs., 10, 172
spinal cord role in, 173–174
Sexual desire, SSRIs and, 156
Sexual differentiation of brain, 21–29. See also Hypothalamus
aromatization theory, 21–22, 24
direct effect of testosterone in, 24–25
factors influencing, 23
and hormone/neurotransmitter metabolism, 26
in humans, 28–29
in hypothalamus and limbic structures, 27–28
Sexual disenfranchisement (spinal cord injury), 139
Sexual identity, 10, 172
Sexuality, complexity of, 172
Sexually dimorphic nucleus (SDN), 35

Sexually dimorphic nucleus of preoptic area (SDN-POA), 30, 31, 33–36, 173
   age-related changes in cell number of, 33
   and copulatory behavior, 27
   development/sexual differentiation of, 34, 38–40
   in humans vs. rats, 36–38
   sex differences in, 32, 172
   sexual differentiation of, 30
Sexual motivation
   and dopamine, 86–87
   and forebrain, 83
   and hypothalamus, 82
   and noradrenaline, 87
Sexual neutrality at birth, 47
Sexual orientation
   amygdala, role in, 28
   and androgen action on brain, 25
   in animals, 16–17
   brain injury and changes in, 151
   diagnosis of, 176
   factors influencing, 23
   following sex change, 172
   genetic factors in and influences on, 50. See Genetics of sexual orientation
   and hypothalmic structure differences, 51–53
   and interaction of developing brain and sex hormones, 39
   measurement of, 9–10
   postnatal social factors in, 51
   prenatal factors in, 50–51
   SDN-POA, role of, 36
   and sex hormones during development, 50–51
   temporal lobe, role of, 28
Sexual rediscovery (spinal cord injury), 139
Sexual reflexes, spinal, 173
Sheep, 16, 28
Sibling studies of sexual orientation, 10–14
Sildenafil, 88
Single-gene hypothesis, 14
Skin (of penis), 78
Social factors in sexual orientation, 50–51

SON. See Supraoptic nucleus
Sphincter reflexes, 81
Spinal cord, 173–174
Spinal cord injury, 109–145
   ejaculation in men with, 79
   orgasm in women with, 79
   sexual response in women with. See Brain-imaging studies
   trajectory of adjustment to, 139
Spinal innervation, 76–83
   forebrain, 83
   hypothalamus, 82–83
   interneurons, 78–79
   medulla and pons, 80–81
   midbrain, 81
   motor mechanisms, 76–77
   sensory mechanisms, 77–78
   spinal reflexes, 79–80
Spinal reflexes, 79–80, 173
SRY gene, 26, 174
SSRIs. See Selective serotonin reuptake inhibitors
Stress, 158–159
   and opiates, 157–158
   and PTSD, 154
   and testosterone, 153
Striated perineal muscles, 77, 79–80
Stroke victims, 151
Subceruleus, 87
Substantia innominata, 49
Suprachiasmatic nucleus (SCN), 30, 31, 39
   and homosexuality, 51–54
   innervated by VIP, 35
   sex differences in, 31
   and sexual orientation, 51
Supraoptic nucleus (SON), 30, 31, 33
   age-related differences in, 43
   oxytocin innervation in, 89
   sex differences in, 42–45
Sympathetic preganglionic neurons, 76, 81

Tactile threshold (in brain-imaging studies), 112, 113
Technologies, research, 175–176

Temporal lobe, 150–152
    change in sexual behavior and
        damage to, 28
    and compulsive sexual behavior,
        173
    sexual-behavior regions of, 173
Testicular steroidogenesis, 162
Testosterone, 24–25, 82
    and CSB, 152–154
    formation of, 21
    measurement of, 153
    and sexual differentiation of brain,
        22, 24
TFMPP (*N*-trifluoromethylphenyl-
    piperazine), 84
Thalamus, 22
Third ventricle, 30, 31
Thornton, L. M., 13
Transneuronal labeling techniques, 176
Transsexuality
    and BST, 45–50, 173
    factors influencing, 23
    incidence of, 45–46
    and prenatal exposure to
        anticonvulsants, 26
    subcategories of, 172
Trauma, 155, 156, 158–160, 163
Trazodone, 85
Twin studies, 10–13
2A receptor subtype, 84, 85
2C receptor subtype, 84, 85

Urinary continence, 79
Uterus, 77

Vagal pathway, 77–78
Vaginal self-stimulation, spinal cord
        injury and. *See* Brain-imaging
        studies
Vaginocervical stimulation, 83
Vagus nerves, 111–115, 135, 137,
        174. *See also* Brain-imaging
        studies
Valproic acid, 161
Van der Kolk, B. A., 155, 158–159
Vasodilatory innervation, 76
Vasopressin
    and erectile function, 44
    sex differences in, 42–45, 54
Ventral tegmental field of midbrain, 81
Ventromedial nucleus of hypothalamus,
        29
Viagra, 88, 177
Viral tracing, 78, 80–83
Viruses, 78
Volatile nitrites, 91

Whitam, F. L., 12
White matter, 76

Xq28 gene, 14, 50

Yohimbine, 87, 88

Zebra finches, genetic studies of, 16–17
ZFY gene, 26

ABOUT THE EDITOR

Janet Shibley Hyde, PhD, is the Helen Thompson Woolley Professor of Psychology and Women's Studies at the University of Wisconsin—Madison. She earned her BA in mathematics from Oberlin College and her PhD in psychology from the University of California, Berkeley. After an early research career as a behavior geneticist, she focused her scholarly energies on the areas of human sexuality and the psychology of women. The author of more than 100 articles and chapters, she has conducted meta-analyses of studies of psychological gender differences in areas such as self-esteem, mathematics performance, and sexuality. Another line of her research has been sexuality in marriage. She is the author of the undergraduate textbook, *Half the Human Experience: The Psychology of Women*, now in its sixth edition, and is coauthor, with John DeLamater, of *Understanding Human Sexuality*, now in it's ninth edition. She has received awards for her research, including the Heritage Award from the Society for the Psychology of Women (American Psychological Association Division 35) and the Kinsey Award from the Society for the Scientific Study of Sexuality.